Design your own Stonehenge - Using the Occam's Razor Solution

By John Hill

Order this book online at www.trafford.com
or email orders@trafford.com

Most Trafford titles are also available at major online book retailers.

Note for Librarians: A cataloguing record for this book is available from Library
and Archives Canada at www.collectionscanada.ca/amicus/index-e.html

Printed in Victoria, BC, Canada.

ISBN: 978-1-4251-9205-1 (sc)

*Our mission is to efficiently provide the world's finest, most comprehensive book publishing
service, enabling every author to experience success. To find out how to publish your book, your
way, and have it available worldwide, visit us online at www.trafford.com*

Trafford rev. 8/3/2009

www.trafford.com

North America & international
toll-free: 1 888 232 4444 (USA & Canada)
phone: 250 383 6864 ♦ fax: 812 355 4082

Table of Contents

List of Tables

List of Figures

Acknowledgements

The author wishes to acknowledge the following people for the tremendous amount of help, support and assistance they provided during the planning and execution of the *2008 Stonehenge Rope Experiment using the principle of Occam's razor.*

The staff and students from the University of Liverpool's School of Archaeology, Classics and Egyptology, in particular Professor Thomas Harrison, Dr. Gina Muskett and Dr. Zosia Archibald. From the University's Sports Directorate, Andy Craig, Peter Sampara and all the ground staff at the University's Wyncote Sports Ground, especially Roger O'Neil. To Brian Coggins, Peter Lloyd and Tommy Thompson for their dedicated assistance with setting out the rope experiment. To Paul Wilson, Diana Duckworth, Gilvan Hamsi Junior and Emma Welsby for acting as tour guides. To the Headmaster, teaching staff and pupils from Northcote Primary School including Roy Morgan, Debbie Kinsella, Jake Summerhayes, Jake Young, Jamie Callow, Aaron Noon, Joseph Dobbie, Charlotte Fairclough, Thomas Baxter, Sophie Calland, Leah Doddenhoff, Jessica Lloyd-Torres and Stephanie Rothwell. To Brian Coggins, Tommy Thompson, Katie Kinsella and John Stewart for their help during the final preparations of this handbook. And last but not least, to my wife Carole and my son Neil, for their unconditional support during all of our very early morning surveys at Stonehenge.

Preface

A brief respite between my part time post graduate research and holding down a full time job has allowed me to document this handbook and publish the theory, methods and results of the *2008 Stonehenge Rope Experiment using the principle of Occam's razor.*

The performance of the rope experiment was, of course, the result of many years of studying, surveying and measuring at both Stonehenge and at a number of its surrounding prehistoric monuments. But the rope experiment was more than just a way of demonstrating how Stonehenge could have been designed using simple methods. When developing the research objectives for the experiment I contemplated deeply about the intellectual capabilities of its original builders. I wanted to push back their conscious awareness of both mathematical number and geometrical shape to a point in their cultural development where neither the thought of number or shape existed as a quantifiable concept. Could such a point in time exist or were the prehistoric builders always capable of counting and performing advanced mathematics?

My idea, and it is only an intuitive one, is that the original builders of Stonehenge might have actually been numerically illiterate and thus the type of numeracy they used when designing Stonehenge was derived from some other form of human ingenuity. I'm thinking here about subconsciously expressing numbers through bodily movement. Say, for instance, during the performance of a primitive tribal dance which requires coordination and timing of an entire group. Thus each individual dancer would have had to coordinate their steps not only to the beat of the drum but also to move in time with the other dancers. It was this ability to harmonise collectively their rhythmic dancing that generated, or evolved into, a form of counting that could be both bodily and visually expressed by way of finger counting. This is the origin of the form of numeracy I proposed to use during the rope experiment. I realise that such an idea is difficult to embrace at first, but surprisingly there are a number of mathematicians who may actually agree with my hypothesis:

"There are a great many unanswered questions relating to the origins of mathematics. It is usually assumed that the subject arose in answer to practical needs. It has been suggested that the art of counting arose in connection with primitive religious ritual and that the ordinal aspect preceded the quantitative concept. In ceremonial rites depicting creation myths it was necessary to call the participants on to the scene in a specific order, and perhaps counting was invented to take care of this problem". (Boyer & Merzbach, 1989:5).

Such was the philosophy behind the theory for the rope experiment. It was an attempt to avoid relying upon all those other anachronistic theories so often associated with the use of advanced mathematics. That Stonehenge could have been built by people who could barely count with their fingers and whose underlying awareness and understanding of numeracy was probably based upon tribal dancing is, no doubt, pushing an idea to its extreme. But let us not forget, Stonehenge was known as the "Giant's Dance" and certainly evidence of tribal dancing has been discovered at the prehistoric barrows such as Winterbourne Whitchurch and Sutton 268 (Burl, 2000:69). Surely then, in order to explain Stonehenge's complex design, using an "Occam's razor", we must think carefully about both the origins and the nature of the mathematics we propose. Moreover, we simply cannot keep on projecting the assumption that these very same prehistoric builders were elite scientists.

John Hill M.A. (Archaeology)
September 2008

1 Introduction

Stonehenge needs no introduction. This prehistoric monument is recognised as both a World Heritage Site and national icon for the United Kingdom. About one million people visit it every year and its world wide appeal is probably un-measurable.

Stonehenge is constantly under archaeological investigation, and yet despite all the advances of modern research we still do not know precisely how a prehistoric, preliterate culture could have both designed and built such a magnificent monument. Indeed, as far back as the 1920's, A. P. Trotter (1927) questioned the capabilities of the "old builders" of Stonehenge: *"What instruments did they use?...surely..that no astronomical knowledge, no calculation, no mathematical instruments, no acquaintance with geometry or surveying was needed"*.

Unfortunately, for almost a hundred years since, satisfactory solutions to Trotter's dilemma have not been possible. But current research being undertaken by the author can present one simple answer to Trotter's dilemma. The author calls this the "Occam's Razor Solution" and he can show how the prehistoric builders of Stonehenge could have designed the monument using no greater scientific or mathematical ability other than being able to count with their fingers, measure the sun's shadow and fold lengths of rope.

This handbook has been published with two specific objectives in mind. The first objective is to provide a summary of the results from the *2008 Stonehenge Rope Experiment using the principle of Occam's razor*. This was a unique archaeological experiment performed at the University of Liverpool's Wyncote Sports Ground, Mather Avenue, Liverpool. The second objective is to present a set of instructions to enable the reader to repeat our rope experiment: which, in other words, means providing you with instructions to build your own Stonehenge. Of course we are not actually recommending that you physically start digging ditches, raising banks and positioning megalithic stones just to test our theory. Although the information presented here would allow you to do just that, we would rather hope that you will be content with simply setting out the monument's ground plan. And, believe it or not, the only "tool box" needed for the job are a couple of lengths of rope, the sun's shadow and your own fingers to count with.

1.1 Occam's razor

Scientists, mathematicians, astronomers and physicists use the principle of Occam's razor when they are faced with a dilemma. The principle relates to the idea that given any number of theories that can equally explain the data (in our case, the various astronomical / mathematical theories associated with designing Stonehenge), then the preferred solution is often the one that involves the least number of variables i.e. the simpler, the more likely the solution (Snow & Brownsberger, 1997: 6-7).

How Stonehenge was designed has long been the subject of many speculative theories. Some say that Stonehenge was designed by builders who possessed knowledge of ancient Greek astronomy and mathematics (which would have meant anticipating such knowledge by almost two thousands years). Others have claimed that Stonehenge was designed to function as an astronomical observatory capable of monitoring the agricultural calendar and even predicting solar and lunar eclipses. However, such theories are not backed up by Stonehenge's actual archaeological record. Furthermore, the prehistoric people who built Stonehenge were a preliterate culture and they have left behind no written records for us to assess their mental capabilities. Thus Stonehenge's architectural design becomes a dilemma – how can we reconcile the monument's design with the mental abilities of a people we know so little about?

The performance of the University of Liverpool's *2008 Stonehenge Rope Experiment using the principle of Occam's razor* was an attempt to demonstrate a method which can explain how a prehistoric culture could have built Stonehenge without the need for its builders to be advanced scientists. Our objective was to replicate Stonehenge's ground plan using the simplest of techniques i.e. folding lengths of rope, measuring the sun's shadow and counting with just our fingers. Ultimately, we achieved our objectives, and thus we have ended up with our demonstrable Occam's Razor Solution.

1.2 Knowing Stonehenge

It must be said, though, that there is no better way of "getting to know" Stonehenge other than setting it out for yourself. (See Figure One). Thus, it is hoped that you the reader will repeat our rope experiment and lay out your own replica of Stonehenge's extant ground plan. Furthermore, we would encourage you to perform such a task as a team or group effort. Indeed, as part of the University's contribution towards the City of Liverpool's 2008 European Capital of Culture celebrations we engaged the help of young pupils from Northcote Primary School (Liverpool) to come along and test out our theory. By doing this our experiment brought together the twin poles of the City's educational community – students from the university working side by side with local primary school pupils. We had the children test our hypothesis that rhythmic movement, such as dancing, could have been incorporated in some manner with the method of finger counting we used to set out our experimental design. The children had fun spinning and dancing with their ropes as they laid out the positions of many of Stonehenge's features. Certainly, they easily mastered the type of mathematics that we proposed the original builders had used (i.e. finger counting). Also, one of the "unexpected" benefits of involving the primary school pupils was to witness the tremendous amount of interest that both archaeology and Stonehenge stimulated in their young minds. Purely as a learning exercise, setting out your own Stonehenge certainly has a lot going for it.

1.3 What's to come in this handbook

It was never the author's intention that this handbook should be published as a scholarly book on Stonehenge, and therefore its style has been, perhaps, written more in the manner of a do-it-yourself or self-assembly instruction book. Section Two presents a brief history of Stonehenge - a complex monument and one that has been subjected to almost 1400 years of use, neglect, change and modification. Section Three discusses the rationale behind the rope experiment's methodology. It explains why we used the three simple methods of folding rope alongside finger-counting numeracy and using the sun's shadow. Section Four presents a report on the rope experiment itself. It starts with listing the research objectives which were declared in advance of the experiment and then ends with a discussion of the major problems encountered. Section Five presents the do-it-yourself instructions for laying out a replica of Stonehenge's extant ground plan. Section Six concludes with a summary of the rope experiment's major findings including its overall result. And finally, it may be worth pointing out that Appendix Three provides a photographic record of the performance of the *2008 Stonehenge Rope Experiment using the principle of Occam's razor*. Readers may wish to view this record prior to reading the handbook itself – quite often a picture can, indeed, tell a thousand words!

1.4 The style of this handbook

The author has given considerable thought as to how this handbook could be used post publication. Like any other do-it-yourself instruction book, he anticipates it being practically followed. Therefore, its practical application has had to take into account how best to present those geometrical schematics he actually used during the rope experiment. Having considered a number of electronic / digital options, the author has decided to reproduce his hand-drawn ink figures in their original format. Whilst the quality of reproducing these figures may suffer somewhat, he still believes that they will provide the best possible guidance.

2 A brief history of Stonehenge

It is far beyond the scope of this handbook to present anything other than the briefest of history with regards to Stonehenge. Readers wishing to find out more about this unique monument's history will find a suitable source of references in the bibliography. Archaeologists have categorised Stonehenge's successive architectural changes by "phases". Although there is no one agreed consensus as to what constitutes the exact sequence of these phases, the following guide is provided by the author as a simple categorisation:

- Phase I refers to a period of time during the Middle Neolithic, between 3000 - 2950 BC.
- Phase II refers to the Later Neolithic period, between 2900 - 2500 BC.
- Phase III refers to the Early Bronze Age, between 2500 – 1600 BC.

3000 to 2950 BC: Phase I. The first phase of Stonehenge began with the construction of what archaeologists today refer to as a henge. This henge earthwork consisted of a circular ditch surrounding an earthen bank which encompassed a large central area. Both the ditch and bank were interrupted by at least two causeway entrances. The widest entrance was orientated towards the north-east with a second, narrower, entrance towards the south. Skirting the central area and located just a few feet inside of the earthen bank was a circuit of evenly spaced post holes that originally contained wooden posts. The positions of these 56 post holes, now referred to as Aubrey Holes (so named after John Aubrey who discovered them in 1666 AD), can still be seen today at Stonehenge marked as they are by the presence of modern concrete rounded-tops.

2900 to 2500 BC: Phase II. Towards the end of the Neolithic, Stonehenge seems to have fluctuated between episodes of use and neglect. The people seemed to have been visiting the "old" henge from time to time and then they appeared to leave it abandoned. Certainly, during this phase, there was evidence of digging and backfilling in parts of the ditch. Also a large number of randomly scattered post holes appeared around parts of the central area and at both entrances. It is unknown what type of structures these post holes supported, but some archaeologists have suggested the possible raising of wooden screens and fences. Incidentally, despite the sporadic visits by the people, Stonehenge became the largest prehistoric cremation cemetery so far found in Britain. It is possible that up to two hundred and fifty cremated remains were deposited over parts of the ditch, bank and central area. Even the empty Aubrey Holes, whose wooden posts had long since rotted away, were used as burial pits for some of the cremated remains.

2650 to 2200 BC: Late Phase II / Early Phase III. During this period, as the transition from the Neolithic into the Early Bronze Age commenced, the first standing stones were raised at Stonehenge. The first stones were the bluestones, which are thought to have been originally transported by either human agency or glaciation from the Preseli hills in South Wales, some one hundred and fifty miles north-west of Stonehenge. These bluestones were initially set out at the centre of the "old" Stonehenge in the formation of two concentric, partial rings. Nothing remains of these rings except for the imprints of their stone holes, which are referred to as the Q and R holes. It is possible that another "foreign" stone from South Wales was also raised at Stonehenge during this period. This was the Altar Stone, which is thought to have originated from Milford Haven.

It is probable that both the Q and R bluestones and the Altar Stone did not stand in their allotted positions for long (perhaps no longer than two hundred years). They were dismantled, no doubt, to make space for the next phase of raising standing stones at Stonehenge - that is, the raising of the great sarsen stones. The chronological sequence for raising each sarsen stone is not clear. Arguably, to geometrically position the various outlying sarsens so that they were concentric with the centre then an obstruction-free area would have been a necessity; therefore it is quite feasible to suggest that the first sarsens raised were the various outlying stones i.e. the Heel Stone, the Slaughter Stone and the four Station Stones (plus a number of other outliers now missing from Stonehenge).

2500 to 2200 BC: Early Phase III. During this period two impressive and unprecedented sarsen stone events occurred at Stonehenge. The first event was the raising of the Outer Sarsen Circle – a ring of thirty huge sarsen stones (with capping lintels) set out in the middle of the "old" henge area in an accurate circle possessing a diameter of 97½ ft. The stones used within this circle were massive, each one weighing approximately thirty tons and being some eighteen feet in length (albeit Stone Number 11 being an exception). It has been estimated that it would have taken the combined effort of about a thousand men seven weeks to manually drag just one of these great stones from their source on the Marlborough Downs, some 20 miles north of Stonehenge (Atkinson, 1986:121). Incidentally, just raising one sarsen stone alone would have been an impressive feat of human ingenuity, but the real masterpiece of engineering was to cap all thirty sarsens with their respective eight ton lintel stones.

The second impressive sarsen event at Stonehenge involved the raising of the five great Trilithons. These Trilithon stones were even larger than the Outer Sarsen Circle stones. In fact, Trilithon Stone number 56 was thirty feet long, making it one of the tallest standing stones found throughout prehistoric Britain. Incidentally, the word Trilithon was coined by the antiquarian William Stukeley from two Greek words *tri* and *lithos* meaning three stones. Thus there were five Trilithons in all (i.e. a combined total of fifteen sarsen stones) set out in the shape of a horseshoe, and they all stood close to the geometrical centre of Stonehenge.

Contemporary with the raising of the great sarsen stones was the widening of Stonehenge's North-east entrance and the appearance of the Avenue. The Avenue is the name given to a prehistoric linear ditch and bank earthwork which functioned as some sort of processional pathway that probably guided the prehistoric people into, and out of, Stonehenge. The Avenue was built in two phases. The first phase of construction dates to about 2400 – 2200 BC and it was attached to Stonehenge's North-east entrance. It has been suggested that the orientation of the Avenue with both Stonehenge's North-east entrance and the position of the Heel Stone was a deliberately chosen act to align all three features with the summer solstice sunrise.

2300 BC to 1800 BC: Middle Phase III. It is difficult to be chronologically precise with regard to the various bluestone arrangements at Stonehenge. Certainly, towards the end of the Early Bronze Age the dismantled Q and R bluestones were re-erected at Stonehenge and their positions do generally correspond to the two extant bluestone formations we see today. The first formation, the Bluestone Circle, probably consisted of up to sixty bluestones standing in a ring with a diameter of 76½ ft. The second bluestone formation, the Bluestone Horseshoe, was initially set out in an oval shaped ring standing just inside of the great Trilithons. Then, after a short period of time, some of the bluestones standing in the north-east quadrant of the oval were removed, leaving behind the extant horseshoe shape.

1800 to 1600 BC: Late Phase III. During this period the final building works took place at Stonehenge. Although no further standing stones were raised, there did appear two concentric rings of holes, referred to as the Y and Z holes, and they were dug just beyond the perimeter of the Outer Sarsen Circle. It's not clear what function these holes actually performed. Certainly the holes themselves would have been large enough to hold standing stones – but no stones were raised. Additionally, during this period the completion of the second, and final, phase of the Avenue began. Initially, the first phase of the Avenue was orientated north-east from Stonehenge but this second phase of modification sees a dramatic swing in its direction and the processional pathway was now diverted south-east for almost two miles eventually terminating at the River Avon.

The Stone-age tool kit

Just what tools did the prehistoric builders use at Stonehenge? Perhaps, we can look for an answer amongst the site's archaeological record. Certainly the artefacts recovered indicate the use of a stone-based technology. For instance, flint axes, sandstone axes, sarsen stone mauls and hammer stones provide us with glimpses of the stone-age tool kit used by the builders. To this tool kit we can add those other perishable items that have failed to survive within the archaeological record i.e. wooden digging sticks, woven baskets to assist with the removal of quarried rubble, ropes (made from either plant-based bast or animal sinew), timbers for rollers, levers and even tree trunks that could have functioned as ladders. Last but not least are the animal bone tools such as the antler bone used as picks and rakes whilst cattle shoulder-blade's were used as shovels. Interestingly, because the age of metal had not yet arrived, there is no evidence of metal tools being used at Stonehenge.

3 Methodology

This section discusses the theory behind the methodology used in the performance of the *2008 Stonehenge Rope Experiment using the principle of Occam's razor.* The methodology itself consisted of three specific techniques:

- The use of rope.
- The use of the sun's shadow.
- The use of "basic" numeracy.

3.1 The use of rope

It must be stated that during the performance of the rope experiment we actually used modern nylon rope. The reason for using this modern rope was based purely on the grounds of safety and practicality. Besides, the experiment was not designed to test the material of the rope itself but rather to demonstrate how it could have been used as a measuring aid to help the prehistoric people design and build Stonehenge. Moreover, it will be noted that there has been no archaeological evidence recovered from Stonehenge to substantiate the claim that rope was ever used at the monument. Unless the conditions are right, rope does not survive well in the archaeological record and it was therefore assumed, as a caveat for the rope experiment, that the prehistoric builders of Stonehenge did actually use rope when designing the monument. However, it can be stated that rope, made from either animal sinew or plant-based bast, was indeed being made and used by the prehistoric people when Stonehenge was being built. For instance, rope dating to 2050 BC was found in association with a tree-trunk at the Norfolk "Seahenge" site at Holme (Pryor, 2001:264); and a wooden bucket containing fragments of rope around its rim was recovered from the Bronze Age Wilsford Shaft, not far from Stonehenge itself (Richards,1996:119). Furthermore, archaeologists such as Professor Richard Atkinson do not deny the possibility that the prehistoric builders could have used rope as a surveying aid when setting out parts of the original Stonehenge (Atkinson, 1986:103).

That rope was the most likely measuring aid used for designing the original Stonehenge was also a conclusion established by the author after performing a number of measuring experiments at Stonehenge. These experiments tested a variety of possible techniques which the prehistoric communities might have used when designing the monument. Of all the techniques tested, rope proved to be the most effective measuring aid. The other alternative aids tested, such as wooden measuring rods, were neither practicable nor efficient in their application.

It was further surmised for the rope experiment that there once lived amongst the prehistoric workforce at Stonehenge particular individuals who possessed specialist "architectural" knowledge for designing monuments. And that such knowledge was integrated into the sequences in which these specialists unravelled their ropes. Specifically in mind here was the idea that the ropes were folded in particular ways for certain monuments - that is, fold the ropes one way for designing a burial chamber, fold it another way for a stone circle. The vision of the prehistoric "religious specialist" comes to mind here and, interestingly, there are a number of ancient cultures that did indeed possess similar rope specialists:

"India, like Egypt, had its "rope-stretchers", and the primitive geometric lore acquired in connection with the laying out of both Hindu temples and altars took the form of a body of knowledge known as Sulvasutras or rules of the cord... The stretching of ropes is strikingly reminiscent of the origin of Egyptian geometry, and its association with temple functions reminds one of the possible ritual origins of mathematics". (Boyer & Merzbach, 1989:233).

The term *Sulvasutras* translates into two words, one, *Sulva* referring to the actual cords used for measuring and, two, *sutra* referring to a book of rules relating to the rituals in which the cords were displayed. The philosophy behind the *Sulvasutras* provides a striking analogy for what we were implying during the rope experiment. That is, that there were similar "religious specialists" capable of organising, supervising and managing the building operations at Stonehenge.

3.2 The use of the sun's shadow

Stonehenge seems to have an unusual relationship with astronomy. Many professional astronomers have attempted to "decode" the secrets of Stonehenge using positional astronomy as the key. Indeed, two distinguished astronomers, Professor Gerald Hawkins (1973) and Professor Fred Hoyle (1977), both proposed that Stonehenge was built to function as an astronomical observatory, and they thought that the monument enabled the prehistoric communities to plan their seasonal calendars (Hawkins, 1973:117), track the complex orbital cycle of the moon around the earth every 18.6 years (Hawkins, 1973:117) and even predict lunar and solar eclipses (Hoyle, 1972:50). If these astronomers are right, then surely Hoyle is correct in his claim that Stonehenge has to be one of the oldest astronomical observatories in the world (Hoyle, 1972:1) – and therefore, the monument's architectural design must have been considerably influenced by the need for it to support "prehistoric astronomy". But not everyone is convinced. Archaeologist Richard Atkinson challenged many of the Stonehenge astronomical theories. According to him, astronomers such as Hawkins did not fully understand the archaeology of Stonehenge, and thus there were too many assumptions in Hawkins' interpretations (Atkinson, 1966). Atkinson also pointed out that all those theories linking the summer solstice alignment with the Heel Stone were flawed. The sun is off centre to the alignment, and it will not be in precise alignment until 3260AD (Atkinson, 1986:30). Another contradiction for the astronomical theories is the fact that archaeologists have yet to recover any artefact from Stonehenge that could provide unequivocal evidence that the monument functioned to the sophisticated levels of astronomy that the likes of Hawkins and Hoyle have proposed (Heggie, 1981:103). This begs the question - does Stonehenge have anything to do with astronomy?

The author suggests that Stonehenge can be designed without involving any complex astronomy. In fact the only astronomical observation he needed to orientate the rope experiment was to measure the sun's shadow at midday. He came upon this observation after examining the positioning of Stonehenge's South entrance in relation to the Outer Sarsen Circle stone Number 11 and the henge's geometrical centre. His conclusion here was that the original positioning and setting out of all three features (even though they are not contemporary) could only have been achieved with practical knowledge of identifying true north. Furthermore, the positioning of the rope experiment's design, using the sun's shadow, could be performed in a matter of days as opposed to alternatively waiting perhaps a full year, or even two, in order to accurately identify the orientations of both the summer and winter solstices. Thus the author had found a quick and simple solution for lying out the rope experiment's design without involving any complex astronomy – and, as the results of the experiment show, it worked.

It is at midday, when the sun reaches its highest point in the sky (referred to as the zenith), that the shadow of, say, a wooden pole, shortens in length and thus indicates the direction of true north. This was the only "astronomical" observation performed during the rope experiment and therefore, perhaps, it might have been the only astronomical alignment needed for setting out the real Stonehenge. However, and again it must be pointed out, there is no archaeological evidence to support the suggestion that the original builders of Stonehenge used the sun's shadow in the manner described here. But it does seem reasonable to assume that knowledge of this universal, age-old practice for determining direction was widely known and used by many other prehistoric cultures (Hogben, 1943:48-51).

3.3 The use of "basic" numeracy

One has to be very careful about proposing whether or not the original builders of Stonehenge were using some kind of mathematical system. The author has already questioned the likes of Hawkins and Hoyle whose astronomical proposals seem anachronistic. And therefore, when investigating for potential mathematics at Stonehenge the author was determined not to make the same anachronistic mistakes as others. Furthermore, there is no archaeological evidence recovered from Stonehenge which could support the idea that its original builders were using any form of mathematics at all. Whatever the method of counting used at Stonehenge, it is most likely to remain a mystery. However, ongoing research conducted by the author, at both Stonehenge and amongst its surrounding prehistoric landscape, has identified a potential form of numeracy and this provided the basis for the mathematics he used during the rope experiment. This form of mathematics can be broken down into its four constituent parts:

- Measuring.
- Finger counting numeracy.
- Units of measurement.
- Geometry.

3.3.1 Measuring

Stonehenge did not stand in isolation upon the Salisbury Plain. In fact, it was built amongst what archaeologists describe as a "ritual landscape". Indeed, this ritual landscape was well under development for several hundred years before the first Stonehenge ditch and bank earthwork appeared. Even today there can still be seen a dozen or so extant Neolithic earthen long barrows and several hundred Bronze Age round barrows all within a five mile radius of Stonehenge. Therefore, according to the assumptions underlying the rope experiment, if the original builders were using lengths of rope as measuring aids when planning and building Stonehenge, then surely they must have been using their measuring techniques at other monuments elsewhere.

Indeed, amongst the dimensions of many of the monuments across this Stonehenge ritual landscape the author has discovered the use of two specific measurements - implying more of a deliberate intention of measuring rather than a fortuitous coincidence of observations. These two measurements are equivalent in length to our modern day imperial measurements of 90 feet and 180 feet and they are found, most especially, amongst the distances between the monuments. But before continuing, it must be stressed here that the author is not implying that the prehistoric builders of Stonehenge were using Imperial measurements. Alternatively the two measurements quoted here could have been expressed, to be more mathematically correct, as "X and ½X". It is simply for ease of explanation that the author continues to use the Imperial standard. Furthermore, it is beyond the scope of this handbook to present a full account regarding the common appearance of both the 90 feet and 180 feet measurements. However, one particular observation can be briefly discussed here as it helps to demonstrate how these two measurements might have been used in both the wider ritual landscape and at Stonehenge itself.

Located about half a mile north of Stonehenge are the barely recognisable remains of a prehistoric linear earthwork known as the North Cursus. This earthwork is considered to date to about 3100 BC (Souden, 1997:46) and it consisted of a set of parallel ditches with internal banks that stretched out across a narrow strip of land. This cursus is just one of many other similar British Neolithic cursuses found elsewhere and it has been noted that they were all laid out with great precision and accuracy, and thus reflect the workmanship of skilful, prehistoric, surveyors (Loveday, 2006). The possible function for these cursuses range from the astronomical to the ritual and the North Cursus at Stonehenge has been interpreted as a processional pathway which, perhaps, guided the prehistoric people through the Neolithic Stonehenge ritual landscape (Darvill, 2006:89). The earthwork itself is orientated east /west and its longest length was measured, according to Robert Newall, at 9090 feet long (Newall, 1959:32). The author has also measured this same strip of land using both Global Positioning Satellite technology and other surveying equipment and has measured its length at 9090 feet (+ /- 5 feet), albeit measuring the extant remains of the earthwork as we find it today. This 9090 feet measurement was just one of many similar dimensions found amongst Stonehenge's surrounding prehistoric ritual landscape and clearly it must have once been a measurement of extreme importance.

Interestingly, it is the mathematical relationship between the 9090 feet length of the North Cursus and the two previous measurements of 90 and 180 feet which is significant. The measurements are proportional. For example, stretch out a 90 feet length of rope 101 times and we have the length of the North Cursus. It seemed therefore both obvious and practical (at least to the author) that the prehistoric people would have used the two "shorter" lengths of rope 90 feet and 180 feet as measuring aids for not only designing their monuments but also for measuring the distances between the monuments. Surely, a rope 9090 feet long would have been too cumbersome for such a use?

3.3.2 Finger counting numeracy

The identification of the two common measurements found amongst the Stonehenge ritual landscape prompted the author to conduct investigations at Stonehenge itself in order to check whether the original builders had used either of the 180 feet and 90 feet lengths of rope. These investigations were also designed to examine whether or not the prehistoric builders had used some form of numeracy that could have been integrated with the folding of the two ropes. However, before the author began his investigations, certain ground rules were laid down. Firstly, it was essential for him to avoid injecting into the investigation process any anachronistic mathematics such as Pythagorean or Euclidean geometry. Secondly, those assistants helping the author were instructed to avoid thinking "mathematically". For instance, they had to avoid using mental arithmetic and also to think carefully about the "number" language they used when discussing the results of their specific measuring tasks – the reason being that the mathematical language we use today originates from an Arabic system which was only introduced into our western culture during the seventh century AD (Delvin, 2000:50). So everyone involved with helping the author tried hard not express verbally words such as "one", "two" or "three", etc, etc. And certainly using pen and paper to add and subtract numbers was not on the agenda either. It is so easy to forget that prehistoric man probably thought, talked and counted in a language so very different from that which we use today. Therefore, the simplest mathematical process that was adopted for the investigations at Stonehenge was finger counting numeracy. As the mathematician Morris Kline so simply puts it:

"Since the process of this method of counting is facilitated by the use of fingers and toes, it is not surprising to learn that primitive man, like a child, used his fingers and toes as a tally to check off the things he counted, in fact, traces of this ancient way of counting are still embedded in our own language today, the word digit meaning not only the numbers 1,2,3… but a finger or toe as well". (Kline, 1977:30).

The 180 feet rope provided the best results at Stonehenge and, indeed, both the decimal and vigesimal system (counting in multiples of ten and twenty respectively) of counting with fingers worked very well with this length of rope. And it was during these same investigations that a fortuitous observation was made. Those people helping the author with his experiments were using their fingers to calculate what measured ropes they needed for each task they worked on. They would automatically count with their fingers like a child pointing and raising one finger to represent one fold of rope, one hand to represent five folds, two hands to represent ten and so on. Furthermore, these same helpers, who by the way were not mathematicians, began to physically report the results of their experiments to the author by way of hand and finger gesturing. In effect, they were using a sign language to visually communicate the mathematics they used. And, as they all agreed, it was far easier to report the results of their tasks in this manner. In fact, they didn't even need to use a verbal language to report their results - it could all be done visually by gesturing with their fingers and shaking lengths of rope. And with a little practice the helpers became very proficient at it. Indeed, one is reminded of the comment raised by the anthropologist Tylor in his survey of primitive counting systems:

"A child learning to count upon its fingers does in a way reproduce a process of the mental history of the human race; that men counted upon their fingers before they found words for the numbers they used… word language not only followed gesture language, but actually grew out of it". (Tylor, 1891:246).

Combining the finger counting numeracy alongside the folding of rope had several outcomes. Firstly, here was a method that could explain how a prehistoric, preliterate culture might have once built Stonehenge. There was no need to write mathematical figures or symbols – it could all be calculated using human fingers. Secondly, the finger counting arithmetic could be visually expressed and communicated by way of sign language – and here was a way to communicate instructions amongst a prehistoric workforce who might not have actually been able to count and or may not have even spoken in the same language.

Of course it can never be proven that the people who built Stonehenge actually counted in the manner proposed by the author. But certainly, ethnographic studies involving finger counting numeracy amongst primitive cultures can provide the basis for both further research and integration with future rope experiments. For instance, the South American Tamanac tribe once used a vigesimal notation system based upon finger and toe counting. The word for five was the same as the word for hand, *amgnaitone*, the word for ten was the same word used to express two hands, *amgna aceponare,* and so on, until the word they used for twenty which meant a man or "one indian", *tevibn itoto,* (Tylor, 1891:247). Such a counting system is an ideal ethnographic analogy that can be easily applied to all future rope experiments.

3.3.3 Units of measurement

The idea that the original builders of Stonehenge might have also folded and / or cut their 180 feet long ropes into shorter specific lengths arose from further analysis of the results discussed above. After examining a number of measurements which we had taken from the dimensions of the stones themselves, it soon became apparent that there was a mathematical ratio between the dimensions of the stones and the 180 feet rope. Moreover, after carefully measuring each stone, the data revealed two common measurements equivalent in length to our Imperial measurements of 3¾ feet and 9 feet. And, it is noted, both these measurements are proportional to the "longer" 180 feet rope (i.e. 180 divided by 3¾ equals 48; 180 divided by 9 equals 20). So here was a link that literally *tied together* the dimensions of the stones at Stonehenge and the 180 feet rope. We refer to these two shorter measurements simply as the "3¾ feet unit of measurement" and the "9 feet unit of measurement".

We concluded that when the prehistoric builders needed precise measurements for specific tasks then they folded and / or cut their 180 feet long ropes into shorter lengths equal to or proportional to either of the two units of measurement. We can only assume that they did this when they needed to "dress" and shape specific stones, and therefore it must have been important for the prehistoric builders to have access to an accurate measuring gauge. Such an idea gave rise to the possibility that some of the stones at Stonehenge may have actually been used as physical measuring gauges against which the builders held their ropes in order to check their measurements. The stones we had in mind for this function were Stonehenge's various outliers which do in fact possess dimensions equal to the two shorter units of measurement i.e. the south facing width of the Heel Stone being equivalent to the 9 feet unit of measurement and the width of Station Stone number 93 being equivalent to the 3 ¾ feet unit of measurement. We thus concluded that this is how the prehistoric builders could obtain their accurate measurements - a conclusion that could explain the high levels of accuracy achieved at Stonehenge. Moreover, we are now led to believe that the real purpose of the outliers, such as the Heel Stone, were to function as measuring gauges.

A few examples as to how the dimensions of some of the stones relate to the two units of measurement can be given. Folding the 180 feet rope twenty times gives us a measurement equal to one "9 feet unit of measurement" (i.e. 180 feet divided by 20 equals 9 feet). The 9 feet unit of measurement is equal to the width of the Heel Stone, a measurement first recorded by both William Stukeley (1740:33) and by John Smith (1771:37). Similarly, Station Stone number 91 is, according to Professor Atkinson (1986:32), 9 feet long or equivalent to one 9 feet unit of measurement; the Slaughter Stone was positioned at a distance of, according to Herbert Stone (1924:118-19), just over 135 feet from the geometrical centre of Stonehenge which is equal to fifteen times the 9 feet unit of measurement and it stood 18 feet above ground level or equal to twice the 9 feet unit of measurement; the average length of an Outer Sarsen Circle stone is, according to Herbert Stone (1924:4), 18 feet long or equal to twice the 9 feet unit of measurement and the estimated average depth of each of its respective stone holes were 4½ feet or equal to half one 9 feet unit of measurement. With regard to the 3¾ feet unit of measurement then we find the average thickness of each Outer Sarsen Circle stone is, according to Herbert Stone (1924:3), 3¾ feet or equal to one 3¾ feet unit of measurement; the average width, at ground level, of each trilithon stone is, according to Herbert Stone (1924:11), 7½ feet or equal to twice the 3¾ feet unit of measurement. A list of further correspondences showing the author's interpretation of the use of both the 9 feet and 3¾ feet units of measurement, alongside comparative measurements published from reliable sources elsewhere, can be found in Appendices One and Two.

3.3.4 The geometry

It is of course very difficult to comment upon the nature of the "geometrical knowledge" possessed by prehistoric people. What shapes dominated their thoughts? Did they consciously conceive in their minds the mental shapes of the circle, square and triangle in the same manner as we can imagine or picture them in our minds today? Such questions may never be answered but this did not stop us from attempting to consider what shapes these people might have actually thought about. Therefore, we attempted to ensure that both the geometry and geometrical shapes we used during our rope experiment were at least contemporary with those shapes found within the respective material culture left behind by the Neolithic and Bronze Age communities. Obviously, without any direct knowledge of their mental geometry to work from, our assumptions were primarily based upon those prehistoric shapes carved into the material of their time.

Contemporary with the Neolithic and Early Bronze Age we see the appearance of what many people refer to as "rock art" or petroglyphs - symbols and shapes that have been carved on a number of prehistoric monuments as well as numerous natural outcrops. Both their meaning and function remain an enigma, and we need not attempt to study or interpret these shapes here beyond, that is, categorising what we can actually see. But we looked at these shapes from a different perspective. We made a special note of those symbols we could not see. The argument being that if we are going to propose an Occam's razor for setting out a replica of Stonehenge's design then our proposed geometry should only include those shapes recorded in the rock art whilst excluding any geometrical shapes not found in the archaeological record.

Stan Beckensall has performed some much needed research with regard to prehistoric rock art (2002). He lists the various geometrical motifs carved into stone that include cups, rings, arcs, lines and chevrons. Similarly, the archaeologist John Barnatt identifies two distinctive patterns of rock art found across the Peak District (Derbyshire) as being cup-marks and the concentric circles (Barnatt & Reeder, 1982). Moreover, from our analysis we deduced a number of shapes that did not appear amongst the Neolithic and Early Bronze Age rock art. Interestingly, we didn't see the shape of the square, the triangle or right-angled triangle. Nor did we see any carvings showing any geometrical awareness of those later complex Platonic designs such as octagons, pentagons or heptagons. Platonic shapes that have recently been suggested as being fundamental to solving the enigma of Stonehenge's design (Johnson, 2008).

It may be difficult for many of us to embrace the idea of a culture that was still yet to consciously grasp or understand, within their minds, geometrical shapes such as the square, right-angle triangle and so on. Arguably, the shape of the circle may have been the one and only important shape needed by the builders of Stonehenge. Therefore, in order for us to present an Occam's razor to Stonehenge's design, then we had to work with only those geometrical shapes we could find to be contemporary with Stonehenge - that is circles, concentric circles, arcs and lines.

4 The 2008 Stonehenge Rope Experiment using the principle of Occam's razor

4.1 Introduction

This section presents a report on the *2008 Stonehenge Rope Experiment using the principle of Occam's razor*. The report will start by reiterating the experiment's research objectives followed by a review of its major highlights.

4.2 Research Objectives

Prior to the performance of the rope experiment a number of research objectives were declared. The rationale behind most of these research objectives centres on the methods used during the rope experiment and as such many of them have already been discussed in Section Three. However, one or two of the other declared objectives will require some further discussion and these will be expanded upon accordingly:

a) The researcher (i.e. the author) will both explain and demonstrate how three "simple" methods can be used to set out an exact replica of Stonehenge's extant ground plan. That is, he will only use lengths of rope, the sun's shadow and finger counting numeracy in order to mark out on a sports field Stonehenge's extant ground plan.

b) The researcher will complete the above objective without using any modern surveying equipment and he will perform the experiment to an accuracy within 6 inches for all comparisons between those Neolithic features measured at the real Stonehenge and those measured during the rope experiment, and within 3 inches for all similar Bronze Age comparisons.

c) The researcher will also lay out the experimental design so that its orientation captures all those astronomical alignments found at the real Stonehenge and yet he will not make reference to any solar or lunar astronomical data at all, other than to measure the sun's shadow at midday.

d) The researcher will explain how he will fold lengths of rope using finger counting numeracy. Furthermore, he will then explain how this numeracy can be visually displayed by physically demonstrating with both his hands and fingers.

e) The researcher will both explain and demonstrate how the lengths of rope used during the rope experiment might have also been subsequently cut into shorter, specific "units of measurement". Moreover, he will then explain how a single, prehistoric "rope specialist" might have used these shorter lengths of rope to organise a variety of work tasks amongst Stonehenge's prehistoric workforce.

The idea behind the second part of objective e) was to explain how a single individual, who will be referred to as the rope specialist, could have both organised and maintained control over a wide range of building tasks at Stonehenge. We have no idea as to the size of the workforce engaged during the building operations associated with Phase III at Stonehenge. However, some archaeologists have speculated, Atkinson proposed that up to 1500 men would have been required to transport all of the eighty-one sarsen stones from the Marlborough Downs to Stonehenge – and it could have taken them five and a half years to do so (Atkinson, 1986:121). Burl proposes that the detailed construction work at Stonehenge itself, that is, the masonry work, the raising of the stones and capping with them with lintels would have taken about 200 men three years to complete (Burl, 2000:370). Arguably the Phase III building operations required a fairly large work force. So how did the rope specialist both organise and manage a large group of workers? The following brief example attempted to explain how the rope specialist did just that. However, in order to keep the explanation simple the author described how the rope specialist could have simultaneously organised the work tasks of a much smaller group i.e. four stonemasons, four carpenters and four manual workers.

The rope specialist takes a length of rope 180 feet long, folds and cuts it into ten equal lengths of 18 feet each. He gives four of these ropes to four stonemasons who he then tasks to go and independently search for four appropriately sized 18 feet long sarsen stones on the Marlborough Downs; the average length of an Outer Sarsen Circle stone being, of course, 18 feet long (Stone, 1924:4). The rope specialist then gives another four 18 feet ropes to four carpenters and tasks each of them to build wooden sledges capable of carrying those four stones from the Marlborough Downs to Stonehenge. In terms of accuracy, both the stonemasons and carpenters are working to the same set of measurements which were given to them by the rope specialist. From the initial ten ropes there are two left over, the rope specialist then takes one of these ropes and both folds and cuts it into four equal lengths of 4¼ feet. He now tasks four manual workers to each dig a stone hole equal in depth to the length of rope i.e. 4½ feet deep; which is, of course, the average depth of an Outer Sarsen Circle stone hole (Stone, 1924:4). Finally, there is one 18 feet rope left over but the rope specialist literally keeps hold of this rope as he uses it to check on the accuracy of the work he has assigned to the others. Some additional examples, based upon the concept of a rope specialist, will be discussed shortly. (See 4.3.2).

f) The researcher will explain how the rope specialist could have used his ropes alongside some form of sign language to communicate his work instructions to a workforce who may not have all spoken in the same language.

It is, of course, noted that we have no idea what language the original builders of Stonehenge were speaking. Nor are we aware that there was a multi-cultural work force engaged in the building activities at Stonehenge and therefore this objective is surmising an assumption. However, neither should it be assumed that everyone was speaking the same language when building the real Stonehenge.

Certainly, there is archaeological evidence of people coming into Britain during the period when the great sarsen stones were being raised. For instance, found not far from Stonehenge was the grave of a middle aged man who has since been dubbed (by the media) the "Amesbury Archer". His skeleton was found under a small round barrow near the village of Amesbury and from strontium analysis of his teeth it was revealed that he had once lived in central Europe before arriving into Britain some time around 2300 BC. And the archer was not alone, he was just one of many other incomers, whom archaeologists refer to as the "Beakers" or the "Beaker Folk": people who were arriving into Britain throughout the Early Bronze Age (2500 – 1600 BC). But did the indigenous British communities speak the same language as the incoming Beakers? Thus the rope experiment attempted to tackle the potential difficulties of people speaking different languages or, at the very least, varied dialects. The proposed solution, similar to the previous objective, was the use of physical gesturing with the ropes alongside a sign language that involved, in some manner, the display of human hands and fingers.

g) The researcher will demonstrate how the ropes can be used to set out the precise positions of all the extant standing stones at Stonehenge.

h) The researcher will demonstrate how the ropes can be used to set out certain complex formations at Stonehenge, such as the Station Stones rectangle and the 56 Aubrey Holes.

The results of the above two objectives will be discussed shortly while their corresponding practical instructions for setting them out will be detailed in Section Five.

i) The researcher will demonstrate how the ropes used for the experiment could have potentially incorporated a body of "secret or mythological knowledge" relating to Stonehenge's architectural design (i.e. similar to the concept of the *Sulvasutras*).

j) The researcher will explain how the measurement for the length of rope 180 feet long, used for setting out the Neolithic features, was determined by reference to other measurements found elsewhere amongst the prehistoric ritual landscape surrounding Stonehenge (e.g. the length of the North Cursus).

The rationale behind the above two objectives has already been explained in Section Three.

k) The researcher will explain and demonstrate why the dimensions found amongst the measurements of the Station Stones rectangle were important and how such measurements subsequently determined the dimensions for both the central stone settings and constructing an earthen ramp capable of raising the sarsen lintel stones.

The above objective will be discussed further within the following report.

4.3 The *2008 Stonehenge Rope Experiment using the principle of Occam's razor* - the report

Although the *2008 Stonehenge Rope Experiment using the principle of Occam's razor* was scheduled to start week commencing 16th June 2008 the author actually performed some preparatory work the week before. This work was vital as it was necessary to ensure that the size of the sports field allocated for the rope experiment was actually large enough to accommodate the design of Stonehenge's ground plan. It is so easy to underestimate the size of land required for marking out Stonehenge's design and it is therefore highly recommended that any reader wishing to follow our instructions should also check that the size of their allotted field is adequate! As a quick guide one will need a flat unobstructed space of at least 360 feet by 360 feet. Furthermore, as the weather conditions were ideal, the author took advantage of identifying the central point from where to start the rope experiment and then measured the sun's shadow at midday in order to establish the direction of true north.

Measuring the sun's shadow is a relatively simple task – but it does require a degree of both patience and time. Burl is correct when he says that the prehistoric people would need two to three days to measure the sun's shadow (Burl, 2000:118) and the author also spent several days performing this task. Although no complex instructions are needed, it was important to ensure that the shadow pole used was capable of both casting a recognisable shadow and stable enough to withstand any breezes. Moreover, the paper the author used to measure the shadow's length was secured firmly to a board and grounded so that it too would not be distorted by any slight breezes. As each morning progressed towards midday the shadow's length was continuously tracked and its position plotted. Once the direction of true north was finally established the author set out a short line of poles in alignment with both the shadow and the direction of true north.

4.3.1 Day One – Monday 16th June 2008

The team of helpers who volunteered to assist the author with setting out the rope experiment all arrived on time and the weather conditions at the start of the week were ideal for marking out. We borrowed a "white line marking machine" from the university's sports ground staff. These machines are very simple to use: basically fill them up with the appropriate white line marker fluid and then just roll them along the areas you want to mark out.

The first features to be marked out were those circles demarcating the Outer-Ditch, the Inner-Bank, the Aubrey Holes circuit and its ring of 56 post holes, the North-east entrance and the South entrance. All the appropriate instructions for setting out these features are included in Section Five and they need not be repeated here. However, it's perhaps worth expanding upon the procedure we used to set out the 56 Aubrey Holes circuit more so because of the scale of astronomical speculation associated with this circuit and, most especially, with the use of the number 56. Much of this speculation has originated from the likes of the astronomers such as Hawkins (1973:174-181) and Hoyle (1977:79-90) both of whom proposed that the Aubrey Holes circuit and its total of 56 post holes were actually used as some kind of prehistoric astronomical eclipse predictor. They claimed that some form of counter was rotated around the circuit of Aubrey Holes using their own versions of numerical formulae which dictated how and where the counter was moved. And when the counter landed upon certain Aubrey Holes then there would be either a solar or lunar eclipse. Thus, according to their reasoning, it was vital that the prehistoric builders of Stonehenge needed 56 Aubrey Holes. Indeed, C.A. Newham also proposed that the number 56 had a special significance associated with the periodic orbital cycles of both the sun and moon i.e. the difference in days between three solar years of 365.242 days and the three lunar years of thirteen lunations each (3 x 29.53 x 13) – (3 x 365.242) being equal to 55.95 (Newham, 1993:28).

Much of this reasoning is of course speculation and without any consideration to the archaeology at Stonehenge. For instance, at what period was the eclipse predictor used? Surely for the system to work properly then an obstruction-free Aubrey Holes circuit would have been required. Yet the Aubrey Holes were both changing in their function and their availability. Initially they started off as a ring containing 56 timber posts (2950 - 2800 BC) - which would have hindered the placement of the counter. Then after the timbers had rotted away the empty post holes were used for cremation burials (2800 – 2650 BC) – burials which might have prohibited the placing of counters for reasons of superstition or whatever. Finally, around 2600 BC four stations stones were erected over the positions of at least four disused Aubrey Holes thus reducing the overall totality of 56 holes. Both Hoyle and Hawkins seem to ignore these factors. Thus, the astronomers seem to start with a given, that is they start with the number 56 and then work backwards trying to figure out why they believe that such a number was significant – and without even considering whether or not there could have been other reasons for choosing the number 56. Alternatively, during our rope experiment, we came up with a total number of 56 Aubrey Holes because that is the number we simply ended up with after folding our ropes and counting with our fingers. Our conclusion is that the number 56 held no specific meaning at all.

The rationale we used for setting out the Aubrey Holes circuit and its 56 holes was to take the length of the rope for the radius of the circuit and fold it using the 9 feet unit of measurement. We had actually measured the radius of the real Aubrey Holes circuit at Stonehenge to be 141¾ feet, incidentally a measurement that is just within a fraction of an inch of Thom's measurement of 141.80 feet (Thom *et al*, 1974:82). So we had a fairly accurate comparison to work from. Now when we divide the 141¾ feet radius of the Aubrey Holes circuit by the 9 feet unit of measurement we get an answer of 15¾ feet, which just happens to be the average distance between each Aubrey Hole. And when we calculate the actual length of the circumference line for a circle with a radius of 141¾ feet we get an answer of 890.64 feet. Thus dividing this length by the average distance between each Aubrey Hole we get 56½. Therefore the maximum number of holes that could be evenly spread out along such a circumference line is 56.

This mathematical theory still works when we use Atkinson's slightly different measurements for both the radius of the Aubrey Holes circuit and the average distance between each hole. He proposed a circuit with a radius of 144 feet and the average distance between each hole of about 16 feet (Atkinson, 1986:27). Thus when we divide a circle with a radius of 144 feet by the 9 feet unit of measurement we get an answer of 16 feet and when we divide the length of the circumference line for such a circle by the average distance between each hole then we still get an answer of 56½ (i.e. 904.77 feet divided by 16 feet equals 56½) and, therefore, the maximum number of holes that could be evenly spread out along such a circumference line, using Atkinson's dimensions, is 56.

However, during the performance of our rope experiment we actually discovered an even easier method (i.e. another Occam's razor) for determining both the radius for the Aubrey Holes circuit and the distance between each hole. After we had set out the central line of the inner bank using the 157½ feet length of rope (See 5.2.5) we folded that rope 10 times or by ten fingers (i.e. into 10 equal lengths of 15¾ feet). We then cut off 1/10ᵗʰ from the rope leaving a residual length of 141¾ feet. We used this residual length of rope to set out the Aubrey Holes circuit and used the discarded 1/10ᵗʰ length of rope (i.e. 15¾ feet long) to set out the average distance between each Aubrey Hole. We started marking out the position of each hole from the southern cardinal point of the Aubrey Holes circuit and began marking each hole accordingly around the circuit, evenly spacing them out at an average distance of 15 ¾ feet. Following a clockwise direction we eventually marked out all 56 holes. Using the conventional numbering system for these holes at the real Stonehenge we started off by marking out Hole No. 21 and then progressed around the circuit until we eventually marked out our final position for Hole No. 22. The final distance between Hole No.21 and Hole No. 22 was measured at 19 feet. Thus, the conclusion from this particular exercise was that despite the importance of the number 56 being stressed by many astronomers – we managed to set out 56 holes without reference to any astronomy at all.

The Station Stones rectangle

Late into Monday afternoon we began marking out the positions of the four Station Stones. These four stones are all positioned equidistant from the geometrical centre of Stonehenge at a distance of 141¾ feet, which is of course a measurement equal to the radius of the Aubrey Holes circuit. When these four stones are linked by straight lines their formation equates to the shape of a rectangle, hence the term the Station Stones rectangle. However, there were no actual features, such as a line of posts or ditches, demarcating this rectangular formation at the real Stonehenge. But their accurate, geometrical positioning in this manner has since tasked the minds of archaeologists. So much so that many scholars think that the rectangular placing of the four stones held significant ritual importance (possibly involving astronomy). Indeed, Atkinson tells us that the setting out of the Station Stones rectangle was *an operation of field geometry which, if it were to be repeated today, would tax the skills of many a professional surveyor* (Atkinson, 1986:33). The procedure we followed for setting out our experimental Station Stones rectangle did not, however, involve any astronomy, and their appropriate instructions are fully explained in Section Five and they need not be repeated here. However, our rectangle did not go quite to plan and before the author explains what went wrong it may be prudent for the reader to familiarise themselves with the geometry involved by studying carefully Figures Eleven and Twelve before reading the following report. (See Figures Eleven and Twelve).

We knew that the four stations stones were positioned upon the same circumference circle line as the Aubrey Hole circuit as it is archaeologically proven that the former stones cut into some of the Aubrey Holes. However, after setting out the positions for the four stones and setting out the theoretical rectangle that connects them we noticed we had an unexpected shortfall in the measurement for the western edge of the rectangle (which would equate to a straight line between Station Stones 93 and 94 at the real Stonehenge). We had to revert to using modern measuring equipment in order to measure precisely what this shortfall was. We actually anticipated this measurement to be about 112½ feet but we had ended up with a measurement of 107 feet – a shortfall of 5½ feet. Initially, we thought a random error had somehow manifested itself into our measurements, most likely as a result of rope stretch. But even after repeating the procedures for setting out Station Stones rectangle we still ended up with a shortfall.

Our rectangle matched a shape which Burl has described as a "regular parallelogram" (Burl, 2000:362). It was at this point that the author recalled Atkinson and Thom's survey of the Station Stones rectangle at Stonehenge (Atkinson, 1978). Indeed our measurements were within acceptable margins of accuracy with their measurements, they too had measured a "parallelogram". (See Table One). Furthermore, we even used GPS to measure the orientations between each of the four corners of our rectangle and compared them with those orientations published by Hawkins (1973:171). (See Table Two). Again we were more or less spot on with our measurements.

Table One – Comparison of Station Stones rectangle measurements between Atkinson's (1978) survey and Hill's Stonehenge rope experiment.

Station Stone	Atkinson's measurement	Hill's Stonehenge rope experiment	Difference
92 to 93	262 feet 3 inches	262 ½ feet	+ 3 inches
94 to 91	263 feet 3 inches	262 ½ feet	– 9 inches
91 to 92	112 feet 10 inches	112 ½ feet	- 4 inches
93 to 91	107 feet 4 inches	107 feet	- 4 inches

Table Two – Comparison of Station Stones rectangle orientations between Hawkins' (1973) azimuth data and Hill's Stonehenge rope experiment.

From	To	Hawkins' Orientations	Hill's Stonehenge Rope Experiment
Station Stone 92	Station Stone 91	049.1 degree	050 degree
Station Stone 93	Station Stone 94	051.5 degree	052 degree
Station Stone 93	Station Stone 92	140.7 degree	142 degree
Station Stone 91	Station Stone 92	229.1 degree	229 degree

Table Two Continued:

Station Stone 94	Station Stone 93	231.5 degree	228 degree
Centre	Station Stone 93	297.4 degree	298 degree
Station Stone 91	Station Stone 94	319.6 degree	321 degree
Centre	Heel Stone	051.3 degree	050 degree

As the first day was coming to a close we concluded that there were two possible explanations for setting out a parallelogram:

- Either, we had followed the exact same procedures for setting out the rectangle as the original builders of Stonehenge had and therefore we were on the right track.
- Or, we had set out the rectangle and somewhere along the sequence of setting out a random error(s) had occurred which resulted with the fortuitous duplication between our measurements and those taken at the real Stonehenge.

The most obvious course of action as this stage would have been to start the rope experiment from the very start and double check every measurement in order to eliminate any random errors, especially those resulting from rope stretch. However, this was, after all, an experiment and who's to say that rope stretch was not a problem also encountered by the original builders of Stonehenge. Moreover, we were also working to a very critical time schedule: our public viewing days were only two days away and there was still a great deal more of marking out to do. Therefore, and for the time being, we marked out on the ground the four straight sides of our parallelogram and any re-examination of our measurements associated with setting out the Station Stones rectangle would have to take place after the public viewing days.

4.3.2 Day Two – Tuesday 17th June 2008

The second day involved the completion of setting out the rope experiment including the marking out of the positions for the Heel Stone and the other central stone features i.e. the Outer Sarsen Circle, the Bluestone Circle, the five Trilithons, the Bluestone Horseshoe and the Altar Stone.

Before marking out the above features, we had to cut some "new" lengths of rope. Firstly, we cut two lengths of rope equal to the lengths of the two longer sides of the Station Stones rectangle, that is two ropes each 262½ feet long. We did this by simply laying down the ropes along the white lines of the longer sides of our rectangle and cutting each rope accordingly. We took one of these ropes, folded it five times (i.e. by counting with one hand or five fingers) and cut it into five equal lengths of 52½ feet long. These five ropes were then referred to as Rope Numbers One, Two, Three, Four and Five which we subsequently used for setting out the remainder of the rope experiment (discussed shortly). The other 262½ feet length of rope was used for setting out the position of the Heel Stone.

Additionally, we had to cut a further two long lengths of rope so that we could link the diagonal corners of our experimental Station Stones rectangle. We needed to link these corners in order to determine the rectangle's geometrical centre. The simplest and most accurate way to do this was to lay out the two ropes from corner to corner and mark the central point where both ropes overlapped. As reported earlier, when we originally set out the Station Stones rectangle on Day One we actually produced a parallelogram rather than a perfect rectangle. So the knock-on effect of this parallelogram was to produce a second geometrical centre from that original centre where we had first started the rope experiment. Our rope experiment now possessed two geometrical centres: one for those Neolithic features which we had previously marked out on Day One and, now, a second centre as a result of our parallelogram. Interestingly the real Stonehenge also possesses two similar geometrical centres, one for the Neolithic features and another for the Bronze Age features (Burl, 2000:366). A summary now follows as to how we used Ropes Numbers One, Two, Three, Four and Five.

Rope Number One

Using Rope Number One, we first stood at our "second" (Bronze Age) geometrical centre and set out a circle with a radius of 52½ feet. After marking out this circle we identified four points around its circumference where the circle crossed the diagonal lines of the rectangle. Using the 52½ feet rope, we set out four arcs from each of these four points. Next we marked the intersections where each arc overlapped and then plotted a straight line linking the overlaps. This new straight line actually represented a new "axis-line" for the orientation of our experiment. Readers may wish to familiarise themselves with Figure Thirteen to help understand this procedure and, although this task may sound complicated, it did not actually cause us any problems and the full set of instructions for marking out this new axis-line are discussed in depth in Section Five (See 5. 2.17 – 18).

After completing the above tasks, Rope Number One was then cut into a number of shorter ropes in order to both demonstrate and explain how the original builders of Stonehenge might have obtained their accurate measurements for both determining and shaping the dimensions of the various stones placed at the monument's centre. The following discussion presents a summary about the sequence we followed:

l) To begin, Rope Number One was folded five times to give five equal lengths of 10½ feet from which we cut off one fifth. We used this discarded one fifth or 10½ feet long rope theoretically to explain how the average length of an Outer Sarsen Circle lintel stone could have been both determined and subsequently dressed to size. The average length of such a lintel stone being 10½ feet long (Stone, 1924:6). Incidentally, the difference between the radius of the inner face of the Outer Sarsen Circle (i.e. 48¾ feet) and the radius of the Bluestone Circle (i.e. 38¼ feet) is also equal to 10½ feet.

m) After cutting off one fifth of the length of Rope Number One we were left with a residual length of rope 42 feet long. From this length we cut off a short measure of rope equal to one 3¾ feet unit of measurement, leaving behind a residual length of 38¼ feet - a measurement that is also equal to the radius of the Bluestone Circle (Stone, 1924:7). The shorter piece of discarded 3¾ feet rope was then used theoretically to explain how the widths of all the Outer Sarsen Circle stones could have been both determined and subsequently dressed to size. The average thickness of such a stone being 3¾ feet (Stone, 1924:3). During the rope experiment the author also explained how he could obtain this 3¾ feet unit of the measurement (and for that matter the 9 feet unit of measurement). Basically, certain outlying stones at Stonehenge were used as measuring gauges where lengths of rope were held against and subsequently cut to size. (See full discussion in Section 3.3.3).

n) Using the 38¼ feet long residual length of Rope Number One we cut off a further length equal to two units of the 3¾ feet unit of measurement (i.e. 7½ feet). This discarded 7½ feet length of rope was then used theoretically to explain how the widths of the great Trilithon stones could have been both determined and subsequently dressed to size. The average width of such a stone (at ground level) being 7½ feet (Stone, 1924:11).

o) By this time the residual length of Rope Number One was now 30¾ feet long and from this rope we cut off a length equal to two 9 feet units of measurement. This discarded 18 feet rope was then used theoretically to explain how the average length of an Outer Sarsen Circle stone could have been both determined and subsequently dressed to size. The average length of such a stone being 18 feet long

(Stone, 1924:4).

p) Rope Number One was now 12¾ feet long and from this rope we cut off a length equal to one half of the 9 feet unit of measurement (i.e. 4½ feet). This discarded 4½ feet rope was then used theoretically to explain how the average depth of an Outer Sarsen Circle stone hole could have been determined and dug to its appropriate depth. The average depth of such a stone hole being 4½ feet (Stone, 1924:4).

q) Finally, we were left with a length of rope 8 feet long. We knew that both the Altar Stone and the Trilithon lintel stones were 16 feet long (Stone, 1924:11-19) so we used our final length of 8 feet theoretically to explain how the dimensions of these stones could have been both determined and subsequently dressed to size. The 8 feet rope being "doubled-up" in its operation.

Rope Numbers Two, Three and Four

These three particular ropes were all simultaneously used to set out the main central features of the experimental design, and none of the associated tasks involved here caused any problems worthy of discussion. Therefore it will not be necessary to repeat their instructions here as they are discussed in depth in Section Five. (See 5.2.20 – 26).

Rope Number Five

Demonstrations using Rope Number Five were not actually physically performed during the rope experiment. Alternatively, explanations were given as to how this particular length of rope could have been used to provide the dimensions to build an earthen ramp capable of raising the lintel stones. The following is a summary of the main points explained by the author during a small mini-exhibition which was run in conjunction with the rope experiment.

The raising of the lintel stones has been the subject of much debate and experimentation. These lintel stones weigh on average eight tons and they had to be raised to a height of at least 13½ feet so that they could cap the Outer Sarsen Circle stones and, indeed, the Trilithon lintel stones had to be raised even higher! Furthermore, the lintels were specifically shaped so that they could sit on top of their respective uprights using a stone copy version of the woodworking techniques for both mortise and tenon alongside tongue and groove. Thus the difficult task of raising the lintels was not just to get them to a specific height - the original builders then had to carefully manoeuvre each one into its precise position - not an easy task to perform with an eight ton stone. There have been many suggestions as to how the original builders raised the lintel stones at Stonehenge. Two popular theories are those proposed by Atkinson and Stone. Atkinson proposed that the lintel stones were raised using timber frames that were laid out in a crib-like structure supporting a flat platform (Atkinson, 1986:134-140). The platform was then raised by constantly levering it upwards by successively underpinning the crib with timbers laid on top of each other. Alternatively, Stone proposed the use of rammed-chalk earthen ramps on which the stones were rolled upwards (similar to the technique we know as parbuckling) to the top of each ramp (Stone, 1924:108-112). Unfortunately, there is no archaeological evidence to support the use of either a timber crib or earthen ramp and therefore whatever method was used at the original Stonehenge will remain a mystery. However, the technique discussed by the author during the rope experiment favoured Stone's earthen ramp theory for the following reasons:

- We could theoretically use our measured ropes alongside our finger counting method of numeracy to provide all the essential mathematical calculations necessary to construct a suitably sized earthen ramp capable of raising an eight ton lintel stone.

- The accuracy required to build such an earthen ramp could easily have been accomplished using one of our ropes 52½ feet long.

- The design of our theoretical earthen ramp not only provided an adequately sized platform capable of handling the intricate manoeuvring of the lintels once they reached their desired height but also the dimensions of the ground plan of the ramp itself would have fitted within the confined space at the real Stonehenge.

- The heights of our theoretical earthen ramp could be easily adjusted using both the 3¾ feet and 9 feet units of measurement.

Stone proposed that for an earthen ramp to be capable of raising an eight ton lintel stone then it would need a slope at an angle of 40 degrees or 5 units vertical to 6 units horizontal. Such a slope would allow Stone's ingenious, yet simple, method of rolling a lintel stone up to its desired height. It is beyond the scope of this paper (and beyond the research objectives of the rope experiment) to review Stone's experiments in any detail; however, it may not be entirely fortuitous that our dimensions obtained from Rope Number Five, with a length of rope 52½ feet long, provided us with the exact dimensions to satisfy Stone's recommended measurements. Folding Rope Number Five ten times (i.e. by ten fingers or two hands) gave us ten equal lengths of 5¼ feet. So for determining the dimensions for the length of the base of our theoretical ramp we multiplied 5¼ feet x six fingers giving a measurement of 31½ feet (incidentally, 31½ feet divided by 3 equals 10½ feet – a measurement that has already been noted elsewhere amongst the dimensions of Stonehenge). We then obtained the desired height for our ramp by multiplying the folded 5¼ feet rope by five fingers which gave us a vertical height of 26¼ feet. We also designed our theoretical ramp to possess a width of 21 feet (i.e. 5¼ x four fingers).

Thus rope No. 5 gave us the theoretical dimensions to construct a suitable earthen ramp with a slope of 40 degrees. Moreover, the overall base area covered by our theoretical ramp was tight enough to fit within the confined working area that would have, no doubt, restricted the building operations at the real Stonehenge. And, finally, we were also able to explain how the height of our ramp could be adjusted by using either of the two units of measurement. For instance, drop the maximum height of our theoretical ramp from 26¼ feet by one 3¾ feet unit of measurement and we end up with a remaining height of 22½ feet which would match closely the same heights, above ground level, for Trilithon Stones numbers 55 and 56 (Stone, 1924:10).

The principle behind using all five ropes simultaneously in the manner proposed by the author was to support the idea that a single individual rope specialist could have actually maintained a level of accuracy across a number of activities - from determining the respective dimensions of each of the various stones (i.e. Rope Number One); to their precise positioning within the central area (i.e. Rope Numbers Two, Three and Four); and for the construction of an appropriately sized, and workable, earthen ramp that would help to place the lintel stones in position (i.e. Rope Number Five). Simply folding one length of 262½ feet rope (a measured length of rope equal to one of the longer sides of the Station Stones rectangle) five times achieved the capability of performing not only a large number of labour-intensive work tasks but it also opened up the possibility that one single person could have planned, designed, managed and maintained accuracy across the entire building operation of Phase III at Stonehenge. Moreover, the logic of the argument used during our rope experiment supports the suggestion that the real function of the Station Stones rectangle at Stonehenge was to provide a controlled working area for determining the dimensions of those features being constructed within the central area. It is in this sense that the real purpose or function of the Station Stones rectangle had nothing to do with astronomy at all.

The Heel Stone

The second 262½ feet long rope was used to set out the position of the Heel Stone and none of the associated tasks involved caused any problems worthy of discussion Therefore, it will not be necessary to repeat these instructions here as they are discussed in Section Five. (See 5.2.19).

5 Instructions for setting out Stonehenge

5.1 Introduction

This section presents the instructions for setting out your own design of Stonehenge's extant ground plan. Before you embark upon such an endeavour, however, it would be prudent to draw your attention to the following points:

Firstly, if you have not already done so, then do read the previous section's report on the performance of the *2008 Stonehenge Rope Experiment using the principle of Occam's razor.* (See 4.3). Secondly, do not underestimate the actual size of the field you will need to set out your own design. To give you some idea of the required amount of space, then a field possessing an unobstructed space of at least 360 feet by 360 feet is essential. Ideally, you should try to get permission to use a sports field belonging to either a school or university and do work closely with their respective grounds staff, especially with regard to their grass cutting schedules. Furthermore request the use of their white line marker machine as this is the recommended tool for marking out your design. Thirdly, do not underestimate the amount of time needed for both setting and marking out your design - allow at least two or three days to complete the entire operation. And finally, if you intend to invite the public to view your completed design, then do ensure that all adequate risk assessments have been taken into account (for instance, do not leave ropes lying unsecured along the ground – people will trip over them!).

5.2 The Sequence

Getting started - setting out the first Neolithic henge earthwork

Your sequence begins with both setting and marking out of those features contemporary with Stonehenge's first henge earthwork, i.e. the Outer-Ditch; the Inner-Bank; the North-east and South entrances; and the Aubrey Holes circuit with its ring of 56 post holes. The diameter of the Outer-Ditch is, according to Professor Anthony Aveni, 360 feet or about 110 meters (Aveni, 2008:87). Dr. Aubrey Burl quotes the same measurement (Burl, 2000:354). Therefore, you will need a length of rope 180 feet long to start with.

Setting out the Outer-Ditch

5.2.1 To begin, identify the central point (position A) from where you will start setting out your design. You will need your 180 feet long rope to start the marking out and, if it helps, you can attach two wooden poles, one to each end of the rope. One person remains in a stationary position at the central point whilst a second person sets out a large circle with a radius of 180 feet which gives you, of course, a circle with a diameter of 360 feet. (See

Figure Two). As the second person sets out the circumference line of this circle, a third person can follow using the white line marker machine. You have now marked out the outer extent of the Outer-Ditch area.

Establishing true north and the north and south cardinal points

5.2.2 The next two procedures involve the identification of true north and the setting out of the north and south cardinal points.

Place a straight pole in the ground at the central point (position A). This post will be used to measure the length of the sun's shadow at midday. To measure the sun's shadow, obviously, a cloudless sky will be required. For best results monitor the sun's shadow from early morning onwards into late afternoon – you may need to perform this task across several days. Mark the moment the shadow reaches its shortest length (i.e. at midday) and then place a line of posts in an alignment with the direction of the short shadow. Initially, this line does not have to be too long, but do use your "line of sight" to keep it straight. You have now established the direction of true north.

5.2.3 Next, use your line of sight to extend the above alignment of posts northwards until the line of posts cuts across the circumference line of the large circle you set out above (in 5.2.1). Place a marker at the north point (position B). At this moment in time you will have a line of posts stretching from position A to position B. Now, you can extend this line directly southwards from position A until the line cuts the circumference of the large circle. Place a marker at the southern position (position E). Providing you have been accurate with using your line of sight you should now have a straight line of posts stretching from the north point (position B) towards the south point (position E), stretching for a distance of 360 feet. (See Figure Three).

Setting out the east and west cardinal points

The next procedure involves the setting out of the east and west cardinal points.

Strictly speaking, to set out these cardinal points you should not use a compass or theodolite. Therefore, in order to determine these points, you will have to rely upon your line of sight and, additionally, some simple manoeuvres with your rope.

One person goes back to the central point (position A) and stands facing position B (i.e. towards true north). This person then raises both arms outwards so that they are at right angles to the north line and parallel to the ground. The right arm indicates the direction of east and the left arm the direction of west. The second person uses the 180 feet rope as a line and stretches it out towards the east in order for the rope to act as a gauge for setting out a straight line of randomly placed posts. Where this post line cuts the circumference line of the large circle place a marker indicating your east cardinal point (position D). Again use your line of sight to keep this line of posts straight. Repeat this procedure for setting out the west cardinal point (position F). You have now set out all four cardinal points. (See Figure Four).

Although the above procedure "crudely" sets out your four cardinal points, you will now have to "tidy up" your measurements so that each point is precisely positioned exactly 180 feet from the central point (position A).

Firstly, remove all the posts that you have been using for setting out the cardinal points. Next, two people are required to hold each end of the 180 feet rope. One person stands at the central point (position A) and the other person stands at the north point (position B). The person standing at position B remains stationary whilst the other person standing at position A holds the rope taut and walks in an "arc-like" direction towards the north-east, they continue walking until they are standing in line with the person at position B as well as forming a right angle with position D; to obtain such a right angle without any equipment, then use the angle between the nose of your own face in relation to your outstretched arms. A temporary marker is now placed at this new position (position C). Now the other person standing at position B walks a south-easterly direction following an arc-like curve towards the east point (position D). If this persons misses (even slightly) position D then the procedure represents either a short-fall or excess in the measured-distance walked and therefore the entire exercise will have to be repeated again until the following four measurements can be achieved i.e. positions A to B are exactly equal 180 feet; positions B to C are equal to 180 feet; positions C to D are equal to 180 feet; and positions D to A are equal 180 feet. (See Figure Five). Once this procedure has been successfully followed it now needs to be repeated for all the other cardinal points. (See Figure Six). It must be stressed that the above exercise cannot be rushed and plenty of time should be allocated to position precisely all your cardinal points in the manner described.

Setting out the North-east entrance

5.2.4 Begin by setting out a line of posts from position A towards position C. Use your line of sight to keep the line straight. Now fold the 180 feet length of rope into 10 equal lengths (i.e. each fold being equal to 18 feet). If you have attached wooden poles to both ends of this rope then now is the time to detach them. Next set out a small circle with a radius equal to one fold of the rope (i.e. 18 feet long) at the point where the line of posts between positions A and C cuts across the circumference line of the large circle you marked out above (as per 5.2.1). You can also set out a second "overlapping" circle upon this A to C line as both these small circles will help to define the width of your entrance. Place markers at the widest points of each of the smaller circles and mark out two straight lines linking them. These straight lines indicate the width of the Neolithic North-east entrance which should be 36 feet wide (See Figure Seven). You have now marked out the North-east entrance.

Setting out the Inner-bank, the widths of the Outer-Ditch
and the South entrance

5.2.5 Now take the 180 feet rope and fold it into 8 equal lengths of 22½ feet. Cut off one fold, that is 1/8th, leaving a residual length of 157½ feet. At this point you may wish to re-attach the wooden poles to both ends of this rope. Two people now use this residual length of rope to set out a circle with a radius of 157½ feet. One person remains stationary at the central position A whilst the other holds the rope taut and sets out the circumference of the circle. A third person follows with the white line marker machine marking out on the ground the circumference line of this circle. You have now marked out the centre line of the Inner-bank and your overall design will, at this moment, possess two large concentric circles on your sports field - one with a radius of 180 feet, the other with a radius of 157½ feet. (See Figure Eight).

5.2.6 Next, use the discarded 1/8th length of rope from above (i.e. 22½ feet long) to set out the widths of the ditch, the bank and the South entrance. Firstly, fold the discarded rope five times (i.e. creating five equal lengths of 4½ feet). Cut off two folds (i.e. equal to 9 feet long) and use this length of rope to set out the width of the Inner-Bank. Use the other length of rope, which should be equal to three folds (i.e. 13½ feet long), to set out the width of the Outer-Ditch. The procedure for setting out the widths of the ditch and bank are simple. Two people hold the respective length of rope taut and walk the respective circumference lines of the two large concentric circles set out above. A third person follows accordingly marking out the lines with the white line marker machine.

To set out the South entrance then the following procedure should be followed. Use the same 13½ feet length of rope above and stand at the south cardinal point (position E). Set out a semicircle along the north / south alignment (as set out in 5.2.3) from the position where the inner edge of the ditch meets the outer edge of the bank. Mark the widest point of the semi-circle (at a 90 degree angle) and set out two straight lines using the 13½ feet rope. Now mark out the South entrance with a width of 13½ feet. (See Figure Nine).

Setting out the Aubrey Holes circuit

5.2.7 Now take the residual length of rope 157½ feet long and fold it 10 times (i.e. ten equal lengths of 15¾ feet). If you have attached wooden poles to both ends of this rope then now is the time to detach them. Cut off one fold leaving a residual length of 141¾ feet and use this rope to set out the Aubrey Hole circuit. At this point you may wish to re-attach the wooden poles to both ends of this rope. One person remains stationary at position A whilst the other holds the rope taut and sets out the circumference of this, now, third concentric circle. A third person can follow using the white line marker machine to mark out on the ground the circumference line of the circle. There are now three large "concentric" circles marked out upon the sports field. (See Figure Ten).

5.2.8 You can use the discarded 1/10th length of rope from above (i.e. 15¾ feet long) to set out the 56 Aubrey Holes. Start setting out your Aubrey Holes from the most southerly point of the circuit, that is, where the circuit's line cuts across the north / south line (between position A and position E) as set out in 5.2.3. Follow a clockwise direction around the circuit and place each Aubrey Hole at an average distance of 15¾ feet apart. You will eventually set out 56 holes BUT you will also end up with a wider gap between your first and last Aubrey Hole position. (See 4.3.1). (See Figure Ten). For marking out each hole, then use a hand held marker spray.

Setting out the Station Stones rectangle

5.2.9 Take the residual 141¾ feet rope used for setting out the Aubrey Holes circuit and fold it in half so that its length is equal to 70.875 feet. Stand at the eastern cardinal point of the Aubrey Holes circuit (position G) and set out a circle with a radius of 70.875 feet. Where the circumference line of this circle cuts across the southern extent of the Aubrey Holes circuit place a marker (position G1). This marker represents the position for Station Stone 91. (See Figure Eleven).

5.2.10 Repeat the same exercise at the western cardinal point of the Aubrey Holes circuit (position H) and where the circumference line of this circle cuts the circumference line of the Aubrey Hole circuit place a marker at the northern overlap of both circles (position H1). This marker now represents the position of Station Stone 93. (See Figure Eleven).

5.2.11 Now, fold the 141¾ feet rope into a quarter that is equal to a length of 35.4375 feet long. Stand at the southern cardinal point of the Aubrey Holes circuit (position J) and set out a circle with a radius of 35.4375 feet. Where both circles overlap place a marker at the eastern overlap (position J1). This overlap marks the position of Station Stone 92. (See Figure Eleven).

5.2.12 Repeat the same exercise at the northern cardinal point of the Aubrey Holes circuit (position I) and where both circles overlap place the next marker at the western overlap (position I1). This marker represents the position of Station Stone 94. (See Figure Eleven).

5.2.13 Now that the four station stones have been positioned you can use your white line marker machine to mark out the Station Stones rectangle (although this is not essential). (See Figure Eleven). However, before you start marking these lines you may wish to consider the following:

The above instructions set out your Station Stones rectangle and providing you have followed both these and all the previous procedures accurately (and you have used a flat, level, sports field) you will end up with a rectangle containing the following approximate dimensions: the two longer sides will be equal to 261 feet and the two shorter sides equal to 112½ feet. However, the reader is reminded that during our *2008 Stonehenge Rope Experiment using the principle of Occam's razor* we encountered a shortfall in the measurements of one of the sides of our rectangle. We actually set out a parallelogram rather than a rectangle. (See 4.3.1). The "shorter" western side of our rectangle turned out to be about five feet less in length than its corresponding (opposite) eastern side. The cause of the shortfall was probably due to random error (as a result of rope stretch) manifesting into the sequence of the procedures we used when setting out our rectangle, but amazingly we still ended up with similar dimensions to the rectangle at the real Stonehenge. This fortuitous result made us question whether the original builders made a deliberate decision to set out the rectangle so that its longer sides equalled 262½ feet. The measurement of 262½ feet is certainly proportional with the 3¾ feet unit of measurement - a unit of measurement that is very prominent amongst the dimensions of the central stone settings and therefore the original builders may have, indeed, deliberately set out a parallelogram.

For those readers setting out their own Stonehenge design then a decision has to be made here. Either carry on regardless with your experimental design or make a deliberate alteration to the dimensions of your Station Stones rectangle. In the overall scheme of things making the following deliberate change alters the overall design of your experiment very little. However, if you do want your final design to replicate Stonehenge's extant ground plan then the following procedure should be followed:

5.2.14 After setting out your Station Stones rectangle as instructed above, a tape measure should now be used to re-position station stones 93 and 94 (referred to as positions H1 and I1). Use your tape measure to measure exactly 262½ feet from position J1 towards position H1, and similarly from position G1 towards position I1. Mark both positions on the appropriate points of the Aubrey Holes circuit. This activity will create two new positions on the Aubrey Holes circuit for stones 93 and 94. (See Figure Twelve).

5.2.15 You will have now set out a parallelogram with the following dimensions – two longer sides of 262½ feet long; one shorter side (the eastern side) with a length of 112½ feet and the other, revised, shorter side (the western side) with a length of 107 feet. At this point you can use the white line marker machine to mark out the rectangle.

5.2.16 Now lay out a length of rope along the full length of one of the longer sides of the Station Stones rectangle and cut it. This length should be 262½ feet long. Next, fold this length of rope five times and cut it into five equal lengths of 52½ feet. There is no fixed rule as to which of these five 52 ½ feet ropes you should now use for the forthcoming instructions, however the author recommends labelling these ropes from numbers one to five and using them in the following manner:

Rope Number One   -Use for instructions 5.2.17; 5.2.18; 5.2.20;
                                   5.2.21.
Rope Number Two   -Use for instruction 5.2.22.
Rope Number Three* -Use for instructions 5.2.23; 5.2.24; 5.2.25;
                                   5.2.26
                                   -5.2.27. * Incidentally Rope Numbers Four
                                   and Five can also be used to assist here.

Re-aligning the axis of Orientation

5.2.17 Now use Rope One (i.e. which is 52½ feet long) so that you can alter the orientation of your design in order to reflect that change of axis which occurred at the real Stonehenge during the Early Bronze Age. This new axis line will be used for orientating both the position of your Heel Stone and for marking out the positions of all the central stone features. To help with understanding the following procedure the reader is encouraged to study carefully Figure Thirteen.

Firstly you will need some additional spare lengths of rope in order to set out two straight lines linking the four diagonal corners of your Station Stones rectangle. If you do not have any spare rope then, alternatively, you could set out two lines of posts between the diagonal corners – but the first option is not only the quickest, it is also the most accurate method. (See Figure Thirteen). By linking the two diagonal corners, this procedure will help you to identify the geometrical central of the rectangle (referred to as position K). You should be aware that if you have followed our recommendation and deliberately altered one of the shorter sides of the Station Stones rectangle (as per 5.2.14) then you will now have a different geometrical centre from the one with which you originally started (i.e. position A).

5.2.18 Next, take Rope One and stand at the geometrical centre of the rectangle (position K) and set out a circle with a radius of 52½ feet. Now walk along the circumference line of this circle and where the circle cuts each of the diagonal lines of the rectangle place a marker at each point (referred to as positions K1, K2, K3 and K4). Now, using Rope One set out four arcs from each of these four markers. Where the arcs generated from positions K1 and K4 overlap (referred to as position K5) place a fifth marker. Where the arcs generated from positions K3 and K2 overlap (referred to as position K6) place a sixth marker. Finally, use your line of sight to mark out a straight line linking positions K, K5 and K6. (See Figure Thirteen).

5.2.19 You can also use the above three positions (i.e. K, K5 and K6) as visual guides for creating another line of posts that extend from the central point (position K) towards the position at which you will place the Heel Stone. Cut an extra length of rope equal to the length of the longer side of the rectangle (i.e. 262½ feet long) and stretch it out in alignment with positions K, K5 and K6 but do remember to lay out the length of rope from the geometrical centre of the rectangle position K. (See Figure Thirteen).

This line now becomes the main "Heel Stone axis line". It is also recommended that you use your white line marker machine to mark out this Heel Stone axis line on your field.

Setting out the central stone features

In order to avoid presenting the reader with too much complex detail regarding the marking out of all the central stone features it will be easier now to make use of a "restored" site plan as devised by Herbert Stone (1924). Whilst Stone's plan may be open to some interpretation it does indeed provide a suitable reference to work from. Furthermore, the author has superimposed his own version of a surveyor's site grid on to Stone's site plan that will, hopefully, help the reader with their next phase of marking out. (See Figure Fourteen). Incidentally, some readers may wish to make their own two dimensional templates of the shapes of the various central stone bases. These templates can then be "cut-out" and subsequently positioned after all marking out has been completed – certainly Stone's site plan will be very useful in this respect.

5.2.20 The first task here starts with setting out the three important focal points along the Heel Stone axis line as well as setting out four major concentric circles. To start, from the central point of the geometrical centre of the Station Stone rectangle (position K) use Rope One and fold it eight times (into equal lengths of 6.5625 feet). Now, standing at position K set out and subsequently mark with the white line marker machine four concentric circles with the following radii:

Concentric circle one = 13.125 feet (i.e. two folds of the 6.5625 feet length)
Concentric circle two = 19.6875 feet (i.e. three folds of the 6.5625 feet length)
Concentric circle three = 26.25 feet (i.e. four folds of the 6.5625 feet length)
Concentric circle four = 32.8125 feet (i.e. five folds of the 6.5625 feet length)

At the point where concentric circle four overlaps the main Heel Stone axis line (see 5.2.19) places two markers. Marker L is placed towards the south-west of position K, marker M is placed towards the north-east of position K. (See Figure Fifteen).

5.2.21 Next, mark out the outer perimeter of the Outer Sarsen Circle. Use your Rope One and stand at position K. Hold the rope taut and set out a circle with a radius of 52½ feet. This circle marks the outer limit of the Outer Sarsen Circle. Now, cut off from this rope a short length equal to one 3¾ feet unit of measurement leaving behind a residual length of 48¾ feet. Now use this 48¾ feet rope to both set out and mark the inner circumference of the Outer Sarsen Circle. (See Figure Fifteen).

5.2.22 You can use Rope Two (i.e. which is 52½ feet long) to set out the positions of the Outer Sarsen Circle lintel stones. The radius of the Outer Sarsen Circle is 52½ feet and interestingly if you divide this measurement by five (i.e. five fingers or one hand) you get an answer of 10 ½ feet which is, of course, the average length of the

Outer Sarsen Circle's lintel stones. Moreover, when you plot the positions of each lintel stone around the circuit you will eventually end up with a final gap equal to 3¾ feet (at the real Stonehenge this 3¾ feet gap is filled by Stone Number 11).

Incidentally, the difference between the radius of the inner face of the Outer Sarsen Circle (which is 48¾ feet) and the radius of the Bluestone Circle (38¼ feet) is also 10½ feet. Obviously, 10½ feet is a significant measurement at Stonehenge.

5.2.23 Setting out Trilithon Stone numbers 51, 52, 59 and 60. Fold Rope Three (i.e. which is 52½ feet long) eight times (i.e. into eight equal lengths of 6.5625 feet). Stand at position M on the Heel Stone axis line and set out four arcs with the following radii:

Arc one = 26.25 feet (equal to four folds)
Arc two = 32.8125 feet (equal to five folds)
Arc three = 39.375 feet (equal to six folds)
Arc four = 45.9375 feet (equal to seven folds)

You can mark out each arc with your white line marker machine but for best visual results keep your markings to within the line set out for concentric circle number four. (See Figure Sixteen). The reader is encouraged to study Figure Sixteen in order to see how each of the four Trilithon Stones were subsequently placed into position. Incidentally, we have recommended setting out the additional arc number four as it helps to define the gap between Trilithon Stones 52 and 59 from Trilithon Stones 53 and 58.

5.2.24 Setting out Trilithon Stone numbers 55, 56, 54, 53, 57 and 58. Fold Rope Three (i.e. which is 52½ feet long) eight times (i.e. into eight equal lengths of 6.5625 feet). Stand at position L on the Heel Stone axis line and set out four arcs with the following radii:

Arc one = 13.125 feet (equal to two folds)
Arc two = 19.6875 feet (equal to three folds)
Arc three = 26.25 feet (equal to four folds)
Arc four = 32.8125 feet (equal to five folds)

You can mark out each arc with your white line marker machine but for best visual results keep your markings to within the line set out for concentric circle number four. (See Figure Seventeen). The reader is encouraged to study carefully Figure Seventeen in order to see how each of the four Trilithon Stones were subsequently placed into position.

If the reader has cut out templates to represent the bases of the Trilithon stones then now is a good time to position them (incidentally, the bases of all the trilithons are equal to 7½ feet by 3¾ feet).

5.2.25 Setting out Bluestone numbers 34, 35, 37, 62, 63, 64, 65, 66, 68, 69 and 70. After setting out the above four arcs (5.2.23) from position L on the Heel Stone axis line set out three additional arcs from this position with the following radii:
Arc five = 39.375 feet (equal to six folds)
Arc six = 45.9375 feet (equal to seven folds)

Arc seven = 52.5 feet (equal to eight folds or one full length of the 52 ½ long rope)

Again you can mark out the line of each arc with your white line marker machine but for best visual results keep your markings to within the line set out for concentric circle number four. (See Figure Eighteen). The reader is encouraged to study carefully Figure Eighteen in order to see how each bluestone was subsequently placed into position.

5.2.26 Setting out Bluestones 61, 31, 32, 33, 46, 47 and 49. Continuing with the series of arcs set out from position M above (see 5.2.23) add three additional arcs with the following radii:

  Arc five = 6.5625 feet (equal to one fold)
Arc six = 13.125 feet (equal to two folds)
Arc seven = 19.6875 feet (equal to three folds)

Again you can mark out each arc line with your white line marker machine but for best visual results keep your markings to within the line set out for concentric circle number four. (See Figure Nineteen). The reader is encouraged to study carefully Figure Nineteen in order to see how each bluestone was subsequently placed into position.

5.2.27 Setting out the "recumbent" Altar Stone. The Altar Stone can be placed where concentric circle number one (set out from position K) overlays arc number three (set out from position L). The reader is encouraged to study carefully Figure Twenty in order to see how the Altar Stone was subsequently placed into position.

Your design of Stonehenge is now complete.

5.3 Validating your results

Once the design of the entire rope experiment has been marked out the reader may wish to validate their results. The author acknowledges that the reader will not be able to freely measure Stonehenge for themselves in order to validate their own measurements. However, Stonehenge is a monument that has been surveyed and measured by many other researchers and a number of these measurements have been published elsewhere. The author has compiled some of these independent measurements alongside his own data (which he did obtain from Stonehenge). Some of this data has been reproduced in two accompanying data tables that should, hopefully, enable the reader to validate some of their own results. (See Appendix One and Appendix Two).

The "best" (i.e. in terms of speed and efficiency) equipment to use to validate your results is a surveyor's 100 metre tape measure (for measuring the radii of the various circles, arcs, etc.) and a Global Positioning Satellite (GPS) handset (for measuring orientation). Such equipment will allow the validation process to be quickly completed within an hour. Regarding your success criteria, the author recommends assessing your results to within an accuracy of 6 inches for all comparisons between your Neolithic measurements and those similar measurements shown in Appendix One and to within 3 inches for all similar Bronze Age comparisons as shown in Appendix Two .

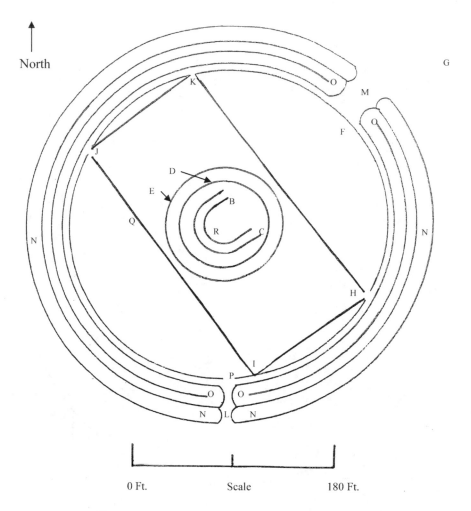

Key

B = Bluestone Horseshoe; C = Sarsen Trilithon; D = Bluestone Circle; E = Outer Sarsen Circle; F = Slaughter Stone; G = Heel Stone; H = Station Stone 91; I = Station Stone 92; J = Station Stone 93; K = Station Stone 94; L = South entrance; M = Northeast entrance; N = Outer-Ditch; O = Inner-Bank; P = Aubrey Holes circuit; Q = Station Stones rectangle; R = Altar Stone.

Figure One - Knowing Stonehenge:
The Author's geometrical plan of Stonehenge's design.

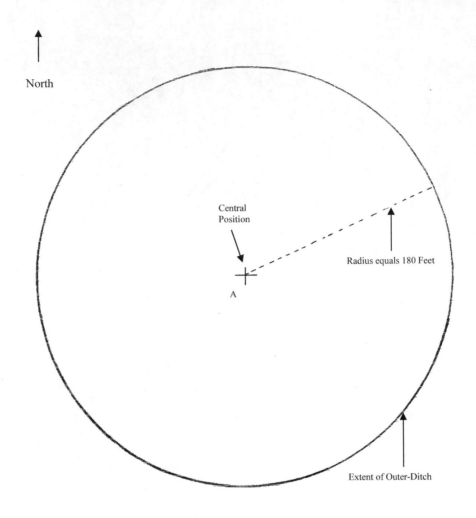

North

Central
Position

Radius equals 180 Feet

A

Extent of Outer-Ditch

Figure Two - Setting the Outer-Ditch

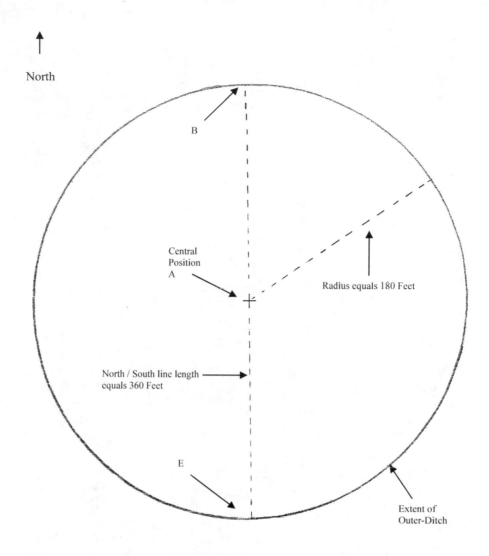

North

B

Central
Position
A

Radius equals 180 Feet

North / South line length
equals 360 Feet

E

Extent of
Outer-Ditch

Figure Three - Setting out of the North/South line

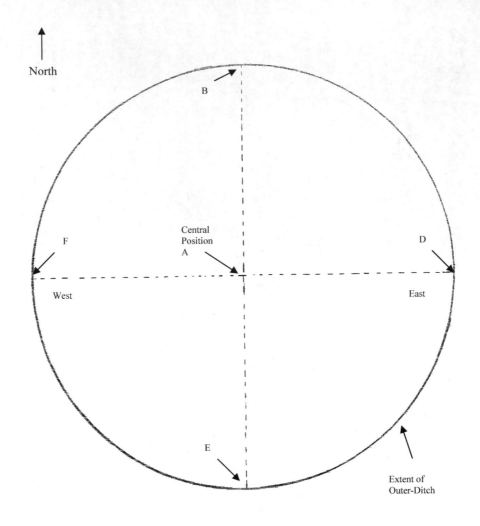

Figure Four - Setting out East & West

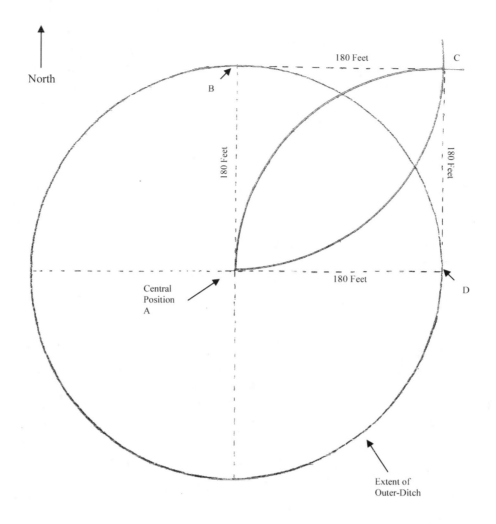

North

B

C

180 Feet

180 Feet

180 Feet

Central
Position
A

180 Feet

D

Extent of
Outer-Ditch

Figure Five - Accurately positioning the cardinal points (part-one)

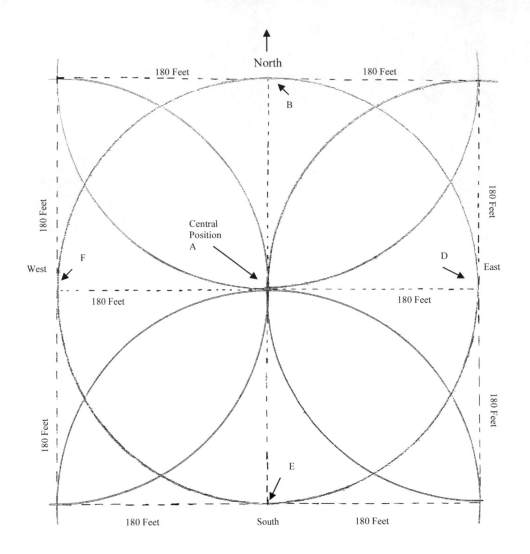

Figure Six - Accurately positioning the cardinal points (part-two)

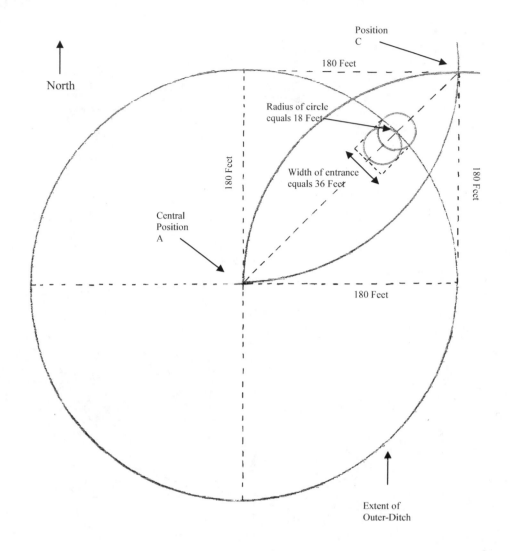

North

Position
C

180 Feet

180 Feet

Radius of circle
equals 18 Feet

180 Feet

Width of entrance
equals 36 Feet

Central
Position
A

180 Feet

Extent of
Outer-Ditch

Figure Seven - Setting out the North-east entrance

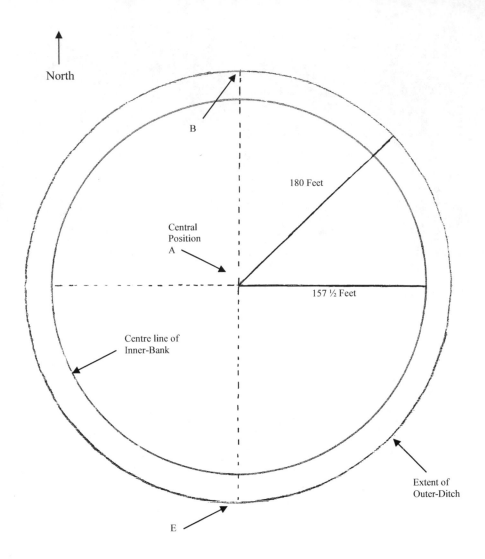

North

B

180 Feet

Central
Position
A

157 ½ Feet

Centre line of
Inner-Bank

Extent of
Outer-Ditch

E

Figure Eight - Setting out the centre line of the Inner-Bank

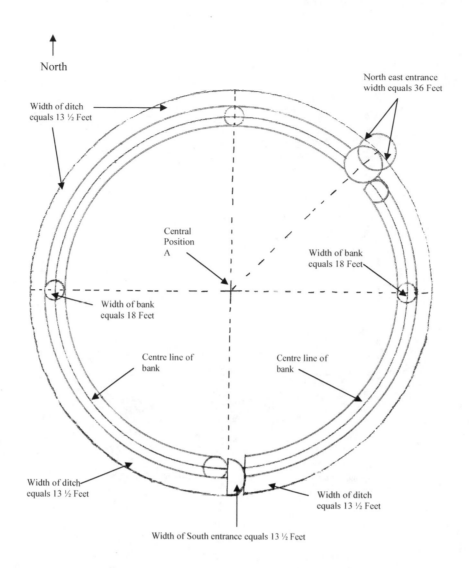

North

North east entrance
width equals 36 Feet

Width of ditch
equals 13 ½ Feet

Central
Position
A

Width of bank
equals 18 Feet

Width of bank
equals 18 Feet

Centre line of
bank

Centre line of
bank

Width of ditch
equals 13 ½ Feet

Width of ditch
equals 13 ½ Feet

Width of South entrance equals 13 ½ Feet

Figure Nine - Setting out the width of the ditch, bank & South entrance

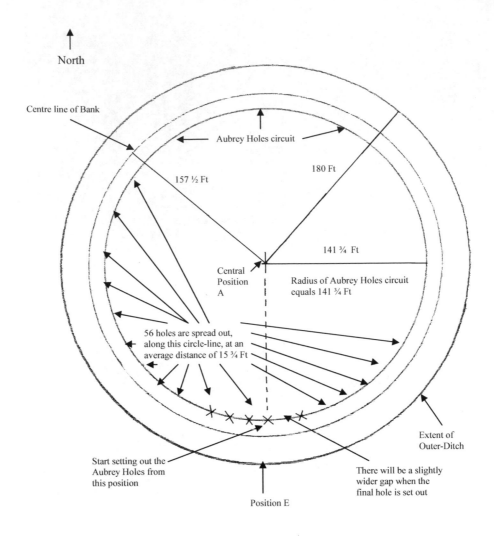

North

Centre line of Bank

Aubrey Holes circuit

157 ½ Ft

180 Ft

141 ¾ Ft

Central
Position
A

Radius of Aubrey Holes circuit
equals 141 ¾ Ft

56 holes are spread out,
along this circle-line, at an
average distance of 15 ¾ Ft

Extent of
Outer-Ditch

Start setting out the
Aubrey Holes from
this position

There will be a slightly
wider gap when the
final hole is set out

Position E

Figure Ten - Setting out the Aubrey Holes circuit

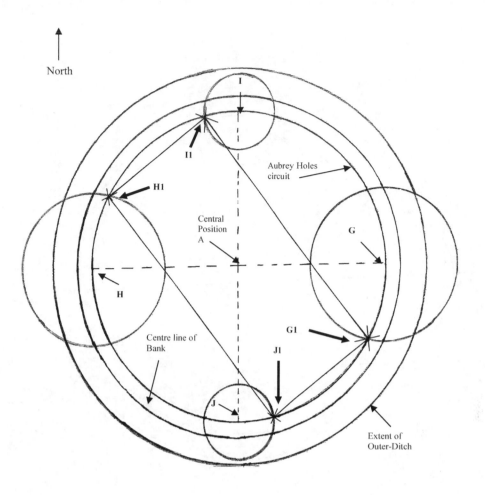

North

I

I1

Aubrey Holes
circuit

H1

Central
Position
A

G

H

Centre line of
Bank

G1

J1

J

Extent of
Outer-Ditch

Figure Eleven - Setting out the Station Stones rectangle

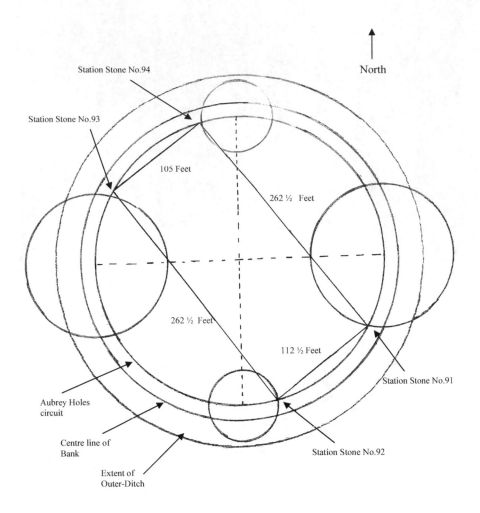

Figure Twelve - Repositioning the Station Stones

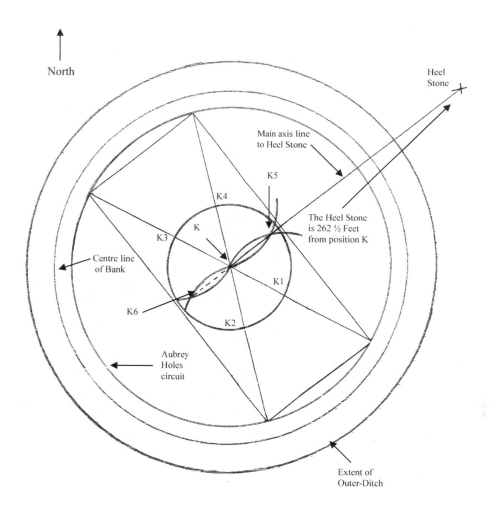

North

Heel
Stone

Main axis line
to Heel Stone

K5

K4

The Heel Stone
is 262 ½ Feet
from position K

K3

K

Centre line
of Bank

K1

K6

K2

Aubrey
Holes
circuit

Extent of
Outer-Ditch

Figure Thirteen - Re-aligning the axis of orientation
(and positioning the Heel Stone)

Geometry of central area at Stonehenge
using restored site plan after Stone (1924) with
superimposed geometrical designs by Hill

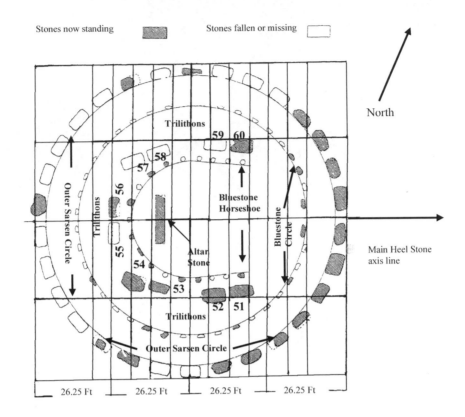

Stones now standing

Stones fallen or missing

North

Trilithons

59 60

58

57

Bluestone
Horseshoe

56

Outer Sarsen Circle

Trilithons

Bluestone Circle

55

Main Heel Stone
axis line

Altar
Stone

54

53

52 51

Trilithons

Outer Sarsen Circle

26.25 Ft   26.25 Ft   26.25 Ft   26.25 Ft

Figure Fourteen – The Central Stone Area
(General Features)

Geometry of central area at Stonehenge
using restored site plan after Stone (1924) with
superimposed geometrical designs by Hill

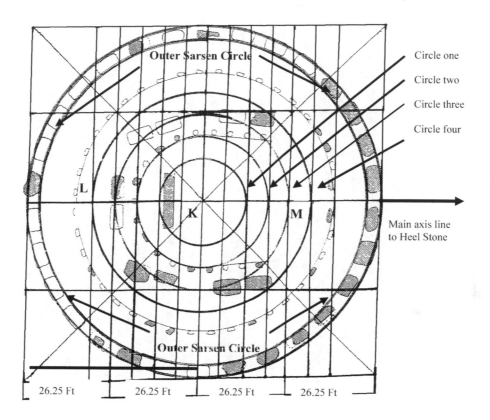

Outer Sarsen Circle

Circle one
Circle two
Circle three
Circle four

L

K

M

Main axis line
to Heel Stone

Outer Sarsen Circle

26.25 Ft   26.25 Ft   26.25 Ft   26.25 Ft

Figure Fifteen - Setting out the four concentric circles &
the Outer Sarsen Circle

Geometry of central area at Stonehenge
using restored site plan after Stone (1924) with
superimposed geometrical designs by Hill

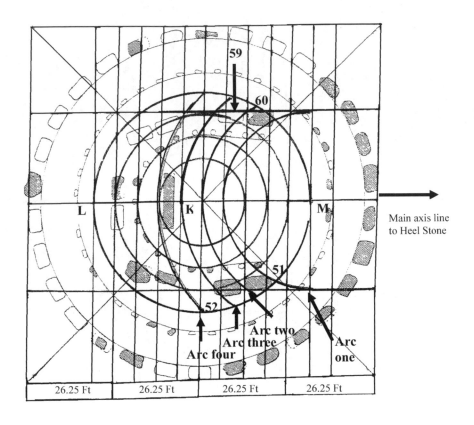

Figure Sixteen - Setting out Trilithon Stones 51, 52, 59 & 60

Geometry of central area at Stonehenge
using restored site plan after Stone (1924) with
superimposed geometrical designs by Hill

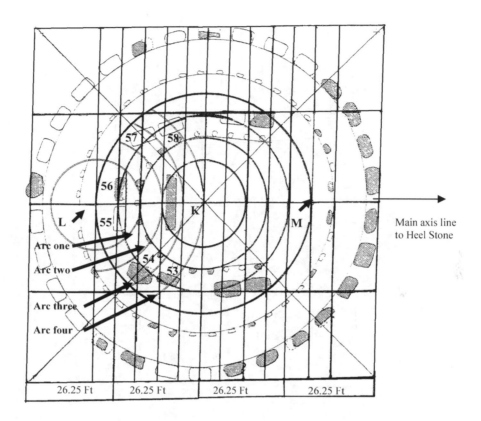

Figure Seventeen - Setting out Trilithon Stones 53, 54, 55, 56, 57 & 58

Geometry of central area at Stonehenge
using restored site plan after Stone (1924) with
superimposed geometrical designs by Hill

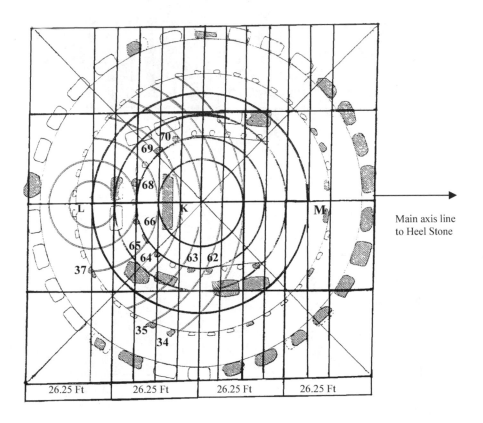

Main axis line
to Heel Stone

| 26.25 Ft | 26.25 Ft | 26.25 Ft | 26.25 Ft |

Figure Eighteen - Setting out the Bluestones (part-one)

Geometry of central area at Stonehenge
using restored site plan after Stone (1924) with
superimposed geometrical designs by Hill

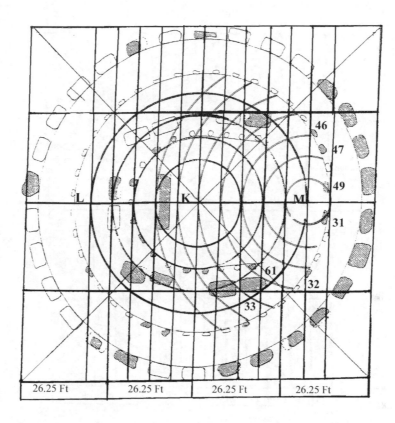

Figure Nineteen - Setting out the Bluestones (part-two)

Geometry of central area at Stonehenge
using restored site plan after Stone (1924) with
superimposed geometrical designs by Hill

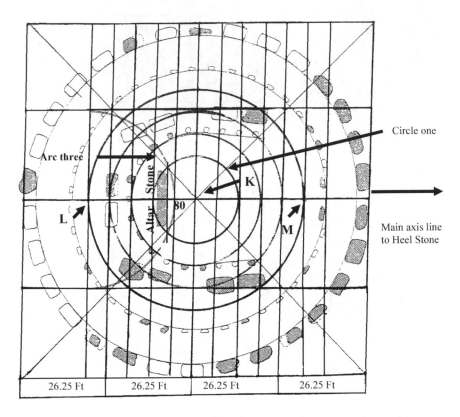

Figure Twenty - Setting out the Altar Stone (stone number 80)

6 Conclusion

Will we ever be able to work out how a prehistoric, preliterate culture could build such a magnificent monument as Stonehenge? For hundreds of years such a quest has been a challenge to the minds of many learned scholars. Perhaps we will never know the answers to this question. Of course, there have been many theories presented as to how Stonehenge was built and, no doubt, the Occam's Razor Solution is just another one. Certainly, the author is not implying that the Occam's Razor Solution is the one and only complete solution. As this handbook has duly noted, there is no archaeological evidence available to support the author's theory that the prehistoric people were using lengths of rope in the manner proposed. But having stated that fact, the Occam's Razor Solution still has a lot going for it, particularly as a working theory that can be practically demonstrated. And perhaps, therefore, it would be prudent to summarise some of its significant findings before considering the experiment's overall result:

- The Occam's Razor Solution (ORS) is an experiment that can be physically demonstrated and independently examined. It produces quantifiable data that can then be assessed and validated against comparative survey data obtained from the real Stonehenge.
- The ORS can set out an accurate replica of Stonehenge's extant ground plan using a combination of "*simple*" methods which are likely to have been known by its original, prehistoric builders i.e. finger counting numeracy, measuring the sun's shadow and folding lengths of rope.
- The ORS can offer a complete sequence of demonstrations to show how the prehistoric builders could have used their measured ropes to set out and position all the major features at Stonehenge i.e. the Outer-Ditch; Inner-Bank; both entrances; the Aubrey Holes circuit; the four Station Stones and all other outlying stones; both the central bluestone and sarsen stone settings and the Altar Stone. Moreover, The ORS can set out all of these features without involving any form of anachronistic astronomy and or mathematics.
- The ORS can be used to explain how Stonehenge's workforce might have been organised. It supports the idea that there once existed, amongst the prehistoric people, individual specialists who could have used their measured ropes not only to design the monument but also organise a variety of work tasks across a workforce who might have spoken in different languages.
- The ORS can be used to provide an explanation of how the prehistoric builders of Stonehenge could have used their measured ropes not only to accurately shape and dress each sarsen stone but also use them to calculate and measure the depths of the corresponding stone holes.
- The ORS can be used to provide an explanation of how the prehistoric builders could have accurately calculated the dimensions of a suitable earthen ramp capable of raising Stonehenge's lintel stones.
- Although not a major claim, it is still notable that the ORS is an experiment whose concept can be both easily understood and practically performed by young primary school children.

6.1 The Result

In terms of assessing the experiment's overall result, the following scoring criteria was produced before the experiment and then subsequently used after all the validation measurements were performed:

The Outer-Ditch & Inner-Bank	10% successful
The North-east & South entrances	10% successful
The Aubrey Holes circuit	10% successful
The Station Stones rectangle	5% partial success
The position of the Outlying stones	10% successful
The Outer Sarsen Circle	10% successful
The Inner Trilithons Setting	10% successful
The Bluestones Circle	10% successful
The Bluestone Horseshoe	10% successful
The position of the Altar Stone	10% successful
Overall Total	95%

Overall, the performance of the *2008 Stonehenge Rope Experiment using the principle of Occam's razor* can be considered a success. However, the author cannot claim a complete 100% success. There was only one procedure which failed to meet the declared objectives and that involved setting out the Station Stones rectangle. (See 4.3.1). Even though we still managed to set out our experimental rectangle within the acceptable levels of accuracy of comparable measurements published elsewhere (Atkinson, 1978) we still believe there was a problem due to a small number of random measuring errors entering into the experiment, especially as a result of rope stretch. And, unfortunately, we will have to wait until we conduct a further rope experiment before any further analysis can take place.

7 Bibliography

Atkinson, R, J, C. 1966. Moonshine on Stonehenge. *Antiquity*, 40, 212-16.

Atkinson, R, J, C. 1978. Some new measurements at Stonehenge. *Nature,* 275, 50-52.

Atkinson, R, J, C. 1986. *Stonehenge.* Middlesex, Penguin Books.

Aveni, A. 2008. *People and the Sky.* London, Thames & Hudson.

Barnatt, J. & Reeder, P. 1982. Prehistoric rock art in the Peak District. *Derbyshire Archaeological Journal,* 102, 33-44.

Beckensall, S. 2002. *Prehistoric Rock Art in Cumbria.* Stroud, Tempus Publishing.

Boyer, C, B. & Merzbach, U, C. 1989. *A History of Mathematics.* New York, John Wiley & Sons.

Burl, A. 2000. *The Stone Circles of Britain, Ireland and Brittany.* London, Yale University Press.

Burl, A. 2006. *Stonehenge – A New History of the World's Greatest Stone Circle.* London, Constable.

Cleal, R, M, J. & Walker, K, E. & Montague, R. 1995. *Stonehenge in its*

Landscape. London, English Heritage.

Darvill, T. 2006. *Stonehenge – The Biography of a Landscape.* Gloucestershire, Tempus Publishing.

Delvin, K. 2000. *The Math Gene.* Great Britain, Weidenfield & Nicolson.

Hawkins, G, S. with White, J, B. 1973. *Stonehenge Decoded.* London, Book Club.

Heggie, D, C. 1981. *Megalithic Science.* London, Thames and Hudson.

Hogben, L. 1943. *Mathematics for the Million.* London, George Allen & Unwin Ltd.

Hoyle, F. 1972. *From Stonehenge to Modern Cosmology.* San Francisco, Freeman Co.

Hoyle, F. 1977. *On Stonehenge.* London, Heinemann Educational Books.

Johnson, A. 2008. *Solving Stonehenge – The new Key to an Ancient Enigma.* London, Thames & Hudson Ltd.

Kline, M. 1977. *Mathematics in Western Culture.* Middlesex, Penguin Books.

Loveday, R. 2006. *Inscribed Across The Landscape – The Cursus Enigma.*

Gloucestershire, Tempus Publishing Ltd.

Newall, R, S. 1959. *Stonehenge.* London, Her Majesty's Stationery Office.

Newham, C, A. 1993. *The Astronomical Significance of Stonehenge.* Wiltshire, Coates & Parker.

Pryor, F. 2001. *Seahenge : New discoveries in Prehistoric Britain.* London, Harper Collins.

Richards, J. 1996. *Stonehenge.* London, B. T. Batsford Ltd.

Smith, J. 1771. *Choir Gaur – Stonehenge on Salisbury Plain.* London.

Snow, T. P. & Brownsberger, K. R.1997. *Universe – Origins and Evolution.* USA, Wadsworth Publishing.

Souden, D. 1997. *Stonehenge – Mysteries of the Stones and Landscape.* London, Collins & Brown Ltd.

Stone, E, H. 1924. *The Stones of Stonehenge.* London, Robert Scott.

Stukeley, W. 1740. *Stonehenge – a Temple Restored to the British*

Druids. London.

Thom, A. & Thom, A, S. & Thom, A, S. 1974. Stonehenge. *Journal History of Astronomy*, 6 (1), 71-90.

Trotter, A, P. 1927. Stonehenge as an Astronomical Instrument. *Antiquity*, 1, 42-53.

Tylor, E, B. 1891. *Primitive Culture – Vol 1*. London, John Murray.

Appendix One - The 9 feet unit of measurement – some comparisons.

Feature	Measurement(s) quoted by other researcher(s)	Proposed measurement using 9 feet unit of measurement during the 2008 Stonehenge Rope Experiment	Difference between comparisons	Actual measurement set out during experiment
The diameter of the Outer-Ditch.	Diameter of outer-ditch equals 110 meters or 360 feet (i.e.180 feet radius). (Burl, 2000:354).	40 x 9 feet unit of measurement equals 360 feet.	Diameter equals Nil.	Diameter of outer-ditch: 360 feet (radius 180 feet).
Width of the Outer-Ditch.	Width of ditch equals 4.2m (13.7 feet) wide (Darvill, 2006:97).	1½ x 9 feet unit of measurement equals 13½ feet.	Diameter equals 2 inches less.	Width: 13½ feet.
Diameter of centre line of Inner-Bank.	Radius 48.3m (i.e.158.4 feet) (Cleal et al, 1995:25).	17½ x 9 feet unit of measurement equals 157½.	9 inches less.	Radius of centre line: 157½ feet.
Width of Inner-Bank.	18.5 feet wide (Cleal et al, 1995:94).	2 x 9 feet unit of measurement equals 18 feet.	6 inches less.	Width: 18 feet.
Aubrey Hole Circuit.	Radius 141.80 feet +/-0.08 ft. (Thom et al, 1974:82).	15¾ x 9 feet unit of measurement equals 141¾ ft.	Less than an inch.	Radius: 141¾ feet.
Average distance between each Aubrey Hole.	"Slightly less than 16 feet apart" (Newham, 1993:23).	1¾ x 9 feet unit of measurement equals 15¾ ft.	3 inches less	Not tested.
Heel Stone.	Width equals 9 feet wide. (Stukeley, 1740:33 & Smith, 1771:37).	1 x 9 feet unit of measurement equals 9 feet.	Nil.	
Heel Stone.	The Heel Stone is located about 256 feet from the centre of Stonehenge. (Stone, 1924:128),	28½ x 9 feet unit of measurement equals 256½ feet.	6 inches more.	Distance from centre to front of stone equals 255¾ feet (9 inches less).

Continued.

Feature	Measurement	9 feet unit of measurement	Error	Result
Slaughter Stone.	The stone stood upright 18 feet above ground.(Stone, 1924:119). It stood about 135.5 feet from centre.(Stone, 1924:119).	2 x 9 feet unit of measurement equals 18 feet. 15 x 9 feet unit of measurement equals 135 feet.	Nil. 6 inches less.	Not tested. Distance from centre: 135 feet.
Station Stones.	The stones are set out on the same circumference of circle of the Aubrey Holes (Cleal et al, 1995:26 & Stone, 1924:114).	15¾ x 9 feet unit of measurement equals 141¾ ft.	Less than an inch.	Radius: 141¾ feet
Station Stone No 91.	Now fallen is 9 feet long. (Atkinson,1986:32).	1 x 9 feet unit of measurement equals 9 feet.	Nil.	Not tested.
Average length of an Outer Sarson Circle Stone. Average depth of each stone hole.	18 feet in length (Stone, 1924:4). 4½ feet (Stone, 1924:4).	2 x 9 feet unit of measurement equals 18 feet. ½ x 9 feet unit of measurement equals 4 ½ feet.	Nil. Nil.	Not tested. Not tested.
Diameter of Inner Bluestone Circle.	76½ feet. (Stone, 1924:7).	8 ½ x 9 feet unit of measurement equals 76½ feet.	Nil.	Diameter: 76½ feet.
North-east Entrance. (Neolithic)	About 36 feet. (Cleal et al, 1995:109).	4 x 9 feet unit of measurement equals 36 feet.	Nil.	Width: 36 feet.
Length of Cursus.	9090 feet long (Newall, 1959:32).	1010 x 9 feet unit of measurement equals 9090 feet.	Nil.	Not tested.

Appendix Two –The 3¾ feet unit of measurement – some comparisons.

Feature	Measurement(s) quoted by other researcher(s)	Proposed measurement using 3¾ feet unit of measurement during the 2008 Stonehenge Rope Experiment	Difference between comparisons	Actual measurement set out during experiment
Station Stones Rectangle.	91 to 92 = 112 ft.10 inches. 92 to 93 = 262 ft. 3 inches. 93 to 94 = 107 ft. 4 inches. 94 to 91 = 263 ft. 3 inches. (Burl, 2006:151).	30 x 3¾ ft. = 112 ft. 6 inches. 30 x 3¾ ft. = 262 ft. 6 inches. 30 x 3¾ ft. = 112 ft 6 inches. 70 x 3¾ ft. = 262 ft. 6 inches.	4 inches less. 3 inches more. 4 ft. 2 inches more. 9 inches less.	91 – 92 = 112 ½ ft. 92 – 93 = 262 ½ ft. 93 – 94 = 107 ft. 94 – 91 = 262 ½ ft. Diameter of circle equals 97½ ft.
Outer Sarsen Circle.	Diameter of circle (inner) equals 97 ½ ft.- radius equals 48¾ ft. (Burl, 2006:30).	Radius therefore equals 13 x 3¾ ft. unit of measurement equals 48.75 ft.	Nil.	
Average thickness of Outer Sarsen Stone.	3¾ ft. (Stone, 1924:3).	1 x 3¾ ft. unit of measurement equals 3¾ ft.	Nil.	Not tested.
Width of Stone 11.	About 4 feet (Stone, 1924:4).	1 x 3¾ ft. unit of measurement equals 3¾ ft.	3 inches less.	Not tested.
Average width of Trilithons at ground level.	7½ ft. (Stone, 1924:11).	2 x 3¾ ft. unit of measurement equals 7½ ft.	Nil.	Not tested.
Bluestone Horseshoe.	The stones in the hollow of the Horseshoe stand in a circle with a diameter of 39 feet (Newall, 1959:10).	10½ x 3¾ ft. unit of measurement equals 39.375 ft.	4 inches more.	Diameter of horseshoe 39.375 feet.

Appendix Three – A Photographic record of the *Stonehenge Rope Experiment using the principle of Occam's razor.*

Photograph 1: Measuring the sun's shadow.

The following series of black and white photographs presents a record of some of the activities performed during the rope experiment. The first task performed was to measure the sun's shadow around the time of midday and establish the direction of true north. This was our only astronomical observation needed in order to orientate the layout of the experimental design. The task itself was relatively simple, but it was very important to ensure that both the recording sheet and the shadow stick could not be distorted by even the slightest of breezes. You may just be able to make out some of the markings we plotted Photograph One.

Photograph 1

Photograph 2: Laying out the first rods.

Once we had identified the direction of true north we laid out a line of rods in alignment with the sun's shadow – each rod casting its own respective shadow on the rod in front of it. In this photograph we can see how both the rods and the sun's shadow were set out pointing towards true north.

Photograph 2

Photograph 3: Extending the true north line.

We used the above alignment of rods and shadow like a compass so that we could correctly orientate an extended line of rods northwards for a distance equal to the 180 feet rope.

Photograph 3

Photographs 4 & 5: Marking out the Outer-Ditch.

In these two photographs we can see the basic procedure involved with both setting and marking out the circumference line of the Neolithic Outer-Ditch. One person holds one end of the rope and remains stationary at the central point, another person holds the other end and sets out the circumference line while a third person follows with the white line marker machine.

Photograph 4

Photograph 5

Photograph 6: Checking for rope stretch.

After each circle was marked out checks were made for rope stretch. This procedure involved making an additional circuit and checking the tension of the rope for any stretching.

Photograph 6

Photograph 7: Setting out the North-east entrance.

We marked out a circle with a diameter of 36 feet at the North-east entrance. We used the diameter of this circle as a guide to determine the width of the entrance (i.e. which is, of course, 36 feet wide).

Photograph 7

Photograph 8: Setting out the centre line of the Inner-Bank.

In this photograph we can see the people both setting and marking out the centre line of the Inner-Bank. This was the second largest circle marked out during the rope experiment. To the outside of this circle can be seen the larger Outer-Ditch circumference line.

Photograph 8

Photograph 9: Marking the Aubrey Holes.

Here we can see how the 56 Aubrey Holes were set out. We used 56 plastic discs to mark out the position of each hole. Then we used a hand held white marker spray to identify each position.

Photograph 9

Photographs 10 & 11: Towards the end of Day One.

In these two photographs we can see the results of Day One's activities. We had marked out all the main circles, set out the 56 Aubrey Holes and marked out the Station Stones rectangle.

Photograph 10

Photograph 11

Photographs 12 & 13: Day Two - The Central Area.

By the end of Day Two we had marked out the entire rope experiment. The following two photographs show the results of Day Two's activities which involved the setting and marking out of the central area.

Photograph 12

Photograph 13

Photographs 14 & 15: Identifying the positions of the various stones.

Just before the public viewing exhibition we used a hand held white marker spray to label the names of the central features. This initiative provided additional clarity during the public tours.

Photograph 14

Photograph 15

Photographs 16 & 17: Pupils test our theory.

We invited pupils from a local primary school to come along and test our theory. The children had fun testing our mathematics whilst also learning about astronomy, archaeology and, of course, Stonehenge.

Photograph 16

Photograph 17

Photograph 18: Build your own Stonehenge.

Running in conjunction with the rope experiment was the Ness Gardens Project. The author was invited by the multi-award winning tourist attraction Ness Gardens (Wirral) to design and build a scaled-down replica model of the first Stonehenge earthwork. The architectural design of this contemporary henge utilised the same three methods employed during the rope experiment (i.e. folding lengths of rope, finger counting numeracy and the sun's shadow). The henge can still be seen today at the Gardens and it imitates that phase of Stonehenge just before the arrival of the great sarsen stones, *circa* 2650 – 2500 BC. Thus it contains all the earlier Neolithic earthwork features alongside the later Heel Stone, the Slaughter Stone and the four Station Stones which, by the way, have all been positioned so that they can observe all the same astronomical observations as found at the real Stonehenge. Moreover, this henge has been orientated towards the distant Welsh mountainous horizon so that every winter solstice the sun sets in the direction of an important prehistoric Bronze and Iron Age site, Moel Arthur.

Photograph 18